THE
ONLY
MATH BOOK
YOU'LL EVER
NEED

Revised Edition

D0954249

THE
ONLY
MATH BOOK
YOU'LL EVER
NEED

Revised Edition

Stanley Kogelman, Ph.D.
and
Barbara R. Heller, M.A.

HarperPerennial

A Division of HarperCollins*Publishers*

Reprinted by arrangement with Facts On File, Inc.

HarperCollins books may be purchased for educational, business, or sales promotional use. For information, please write: Special Markets Department, HarperCollins Publishers, Inc., 10 East 53rd Street, New York, NY 10022.

FIRST HARPERPERENNIAL EDITION

The Library of Congress has catalogued the hardcover edition as follows:

Kogelman, Stanley.
    The only math book you'll ever need / Stanley Kogelman and Barbara R. Heller.
    —Rev. ed.
        p.   cm.
    Includes index.
    ISBN 0-8160-2767-6
    1. Mathematics. I. Heller, Barbara R. (Barbara Rita), 1933–   . II. Title.
QA39.2.K63   1993
513—dc20   93-10539

A British CIP catalogue record for this book is available from the British Library.

ISBN 0-06-272507-6 (pbk.)

        01 00   RRD   19 20

Contents

Preface to the First Edition

The first time we talked about this book was in the airport, waiting for our flight to Rochester, New York, where we were giving a workshop for college mathematics professors on how to teach math in a nonanxiety-producing way.

We had been conducting these workshops at various college campuses for several years, because we believed that people's dislike of mathematics stemmed largely from the way it had been taught to them—with lots of rigid rules that did not allow for creative approaches to the solution of problems—and from the fact that it dealt with seemingly irrelevant problems, like coins and "distances."

We were a good workshop team. Stan, who is extremely math competent, felt strongly that mathematics was not only an important skill, but had a logic and internal elegance that made solving problems fun. Barbara, while not exactly a math pro, works with budgets, statistics, and computers everyday at home and at work. She agreed that mathematics is a *necessary* skill, but she was not convinced it could be enjoyable.

Our back-and-forth discussions about math and about the way typical students learn it stimulated the teachers in the workshops to question their classroom methods and their expectations for students. As a result of this exchange, all of them became more attuned to students' reactions; many changed their teaching techniques.

However, the workshops also made it clear that there were no math materials for adults. Even with new teaching strategies, teachers rarely applied them to the daily kinds of situations involving mathematics that most adults face. Which brings us to that day in the airport.

We began to talk about a book that would address the mathematics in real-life situations, and that would be written in such a way as to balance Stan's intuitive facility with math with Barbara's need to understand each of the sequential underlying steps.

Some time has elapsed between our initial idea and the day that *The Only Math Book You'll Ever Need* went to press for the first time. In that interval, as a result of doing the background research and of talking about the book to friends, colleagues, and acquaintances, the book changed character, developing a somewhat different emphasis. It still deals with adult issues—math for finances, recreational math, and the kind of math you need around the house—and it's filled with facts and ideas that we found interesting. But we played down the theory considerably. While we still showed all the mathematical steps you need to follow, we rarely explained anything general about decimals, percentages, or fractions. Writing about math, just like *doing* math, requires a great amount of patience and practice—and a great amount of space. You will find many places in this book where a seemingly simple example or statement takes several pages to explain step by step.

We hope that after reading this book you will be more open to approaching math-based problems; more willing to attempt the math yourself, and more likely to undertake a greater analysis of your own affairs, thereby gaining an added measure of control. More control is really what it's all about.

Stanely Kogelman
Barbara R. Heller
October 1986

Preface to the Revised Edition

In the interim between 1986 and the publication of this revised edition of the book, there have been some important changes in the way mathematics is being taught (in the elementary grades in particular), and in the kinds of materials and activities that are being made available to young people. One of the critical things we noted, however, was that, based on a March 1992 study funded in part by the Exxon Education Foundation, one of the biggest problems in reforming the teaching of math remains the lack of good materials. So, for a long time, we talked about the changes we needed to make in this edition. We reviewed the letters and notes we received from readers, spoke with our editor, and asked family, friends, and colleagues for advice and suggestions.

The original book was *useful*. In the end, we decided to keep *all* of the material in that original version, but to update it so both the facts were current and the ideas about how to do math reflected the newest thinking. Thus, we've included the most recent income tax schedules in Chapter 1, Section 3 and, because of inflation, increased the cost of a perm (Chapter 2, Section 1), supermarket prices (Chapter 5, Section 1), and the salaries of workers at the Freehold Company (Chapter 9, Section 2)—to mention only a few of these kinds of changes. We also placed more emphasis on mental arithmetic by adding more examples of ways to make it easy to figure things out in your head. This is just one of the new ways mathematics is now being taught.

We also wrote a great amount of *new* material and completely rewrote other chapters. As an illustration, take a look at the sections on retirement annuities and zero coupon bonds and the chapter on computers. These additions reflect the interests of our audience as well as how mathematics continues to enter our lives.

To make this revision even more useful, perhaps the most significant change is the incorporation of ✓TIPS and ✓✓HOT TIPS. Throughout this book, we have highlighted many ways to estimate and approximate. Each ✓TIP–✓✓HOT TIP is based on solid mathematical understanding and lets you arrive at quick, shortcut answers to problems and questions. Note that each ✓TIP also gives you an answer that is not *exact*, but that, in many situations, is all you need to know. In other instances, the ✓TIP–✓✓HOT TIP method of figuring out the answer gives you a good first approximation; then, you might want to calculate the exact amount. Much to Barbara's delight, Stan's shortcuts work all of the time—and they really are fun to use.

So, go ahead, use this book, ✓TIPS, ✓✓HOT TIPS, and all.

Stanley Kogelman
Barbara R. Heller
March 1993

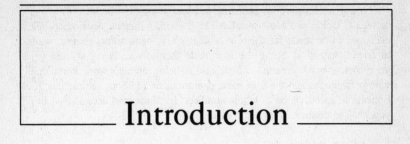

Introduction

In buying this book, you engaged in a financial transaction. It also required a mathematical operation—subtraction—to give you the correct change, plus percentages, multiplication, and addition if you live in a place that has a sales tax.

Every time you spend money, make change, or use your credit card, you are presented with some aspect of everyday math.

The mathematics of money calls for a range of skills. There's subtraction for computing change, and percentages when you need to figure out the amount of tip or sales tax. Percentage increases (or decreases) come into play when you are dealing with inflation; exponents underlie the compound interest formula that goes into decisions about borrowing or investing money; statistics are involved in budgeting; and logarithms are a necessity in computing how long it takes to reach your investment goals.

While financial matters may make the most critical demands on your skills, spending money is not the only daily activity that involves math. Mathematics pervades things people do for fun, such as driving a sports car, traveling abroad, or taking photographs. Building a fence requires an understanding of "perimeter"; the concept of "area" comes in when you carpet your den; recipes need to be doubled or halved; and sewing a round tablecloth makes use of the geometry of a circle.

Math is front page news.

Examples of everyday math can be found in the daily newspaper: in advertisers' claims as well as in business and economic news analyses, sports statistics, science reports, and lottery games. And this doesn't begin to count the math that people need at work or even at home!

Many adults give up choosing the *best* alternative because they can't, or won't, tackle problems that have to do with numbers. Still others allow decisions to be made for them by a sales clerk, bank teller, spouse, waiter, or friend instead of doing the arithmetic themselves. Being willing to do your own everyday math—which means being able to work comfortably with percentages and the four basic operations of addition, subtraction, multiplication, and division of whole numbers, fractions, and decimals—can put you back in control. Being in control:

- Saves you money
- Makes you a more powerful consumer
- Boosts your confidence
- Lets you have more fun
- Expands your options
- Reassures you
- Makes you feel good about yourself.

It may even make you a better cook.

The math you need to be self-reliant is the same math you were taught, and supposedly learned, in elementary school. But most of us were not taught how to apply math concepts to practical, adult life situations. Classroom math was taught—and unfortunately, in many instances, continues to be taught—so that children can learn *more* classroom math. The basic concepts and operations were introduced and are still being taught, by grade school teachers—many of whom have not themselves mastered math's fundamentals and thus teach it poorly, devoting insufficient class time to it.

The National Science Board Commission on Precollege Education and Mathematics, Science and Technology reported that:

"Many of the teachers in elementary schools are not qualified to teach math . . . for even 30 minutes a day.

"A significant fraction of . . . secondary school teachers are called upon to work in subjects for which they were never trained.

". . . there are currently severe shortages of qualified math . . . teachers in many parts of the Nation. Fewer college students are entering the teaching profession, particularly math and science teaching, and increasing numbers of experienced teachers . . . are leaving."

The situation is not yet dramatically improved and the shortage of math and science teachers is projected to continue well into the next century.

The 1990 National Assessment of Educational Progress (NAEP) found that most fourth graders were still being taught by teachers who were not certified in mathematics and who reported little or no course work in the subject. Nonetheless, the nation's students made significant gains on mathematics achievement between 1990 and 1992: overall proficiency rose five points in the 4th, 8th, and 12th grades—although 37 percent of the students failed to reach what is considered a *basic* level of achievement. While American students are improving in math, they still lag behind students from other industrialized nations, according to a study reported in the January 1993 issue of *Science* magazine.

Largely because of inadequate teacher training and preparation, math is taught by means of rules and regulations that are neither correct nor helpful to the student. And it continues to be taught by insecure professionals who, unfortunately, introduce guilt, dislike, and anxiety into the subject. "Ignorance in . . . math teachers begets ignorance in students," the National Academy of Sciences concluded in an important report released in 1991.

It's hardly surprising, then, that a majority of young people grow up with poor skills and uneasy feelings about math. In some cases the negative feelings become so extreme that otherwise capable, intelligent adults can hardly do math at all.

Most young students stop taking math courses as soon as they are free to make the choice. The National Science Board Commission found that, "since the late 1960s, most students have taken fewer mathematics courses. Mathematics . . . achievement scores of 17-year-olds have dropped steadily and dramatically" from the 1960s to the 1980s. The NAEP study conducted in 1992 showed some gains for these 12th graders, however.

In the late 1960s, "62 percent (of U.S. students) did not take Algebra II (in high school), 48 percent did not take geometry . . . Algebra I enrollments fell to 64 percent in 1981 from 76 percent in 1969." By 1990, however, only 9 percent of high school seniors reported having never taken algebra, while the majority (55%) had taken geometry. Girls still lagged behind boys, and African Americans and Hispanics scored below whites in both enrollments and achievement in math courses.

Since coping with today's world requires more math than ever, our schools should continue to require *more* math and better trained teachers. The world abounds with statistics, calculators, computers. It has been estimated that at least 90 percent (that's nine out of 10!) of all jobs today involve computers directly or indirectly. Home computers number in the millions. Even the choice of a savings account, once a simple decision, now presents complex options, including variations in interest rates, compounding periods, and length of time of deposit. Greater earnings are available to the person who takes the time to figure out the alternatives. And that's only one benefit to be gained from mastering basic math.

The Only Math Book You'll Ever Need helps you solve practical financial problems: balancing your checkbook, comparing investment alternatives, and determining mortgage payments. It shows you how to convert foreign currency and offers aids for cooking and sewing. It fills in the gap between what you were once taught and what you need to know now.

Because we want this book to be useful, we show you *what to do,* escorting you through the mathematics so you can compute for yourself the amount of paint you need for an $8'5'' \times 12'6''$ room or how much interest you have to pay if you charge the new raincoat. And we include both short-cuts—✓TIPS and ✓✓HOT TIPS—such as how to estimate when the exact amount doesn't matter, and realistic encouragement, such as "Count on your fingers if it helps you." Many of us are secret finger-counters anyway, despite the repeated admonishments of teachers and parents.

The Only Math Book You'll Ever Need is a collaborative effort of two people whose combined experience totals more than 60 years (that's statistics for you!).

Dr. Stanley Kogelman was chairman of the Mathematics Department at the State University of New York at Purchase for five years. He spent 20 years as a classroom teacher, covering everything from remedial math for adults to math-beyond-calculus for mathematics majors. Dr. Kogelman has taught regular mathematics courses and in special programs, working with individuals and small groups, and with students, prospective teachers, and teachers-as-students. His consulting company, "Mind Over Math," and book of the same name summarized some of his experiences with learners of all kinds and brought him national prominence as a champion of the math-anxious person. Dr. Kogelman has been instrumental in defining how attitudes and feelings affect learning and in proposing techniques to ameliorate negative attitudes toward learning.

A Ph.D. in mathematics and the holder of an M.S.W. degree, he also worked as a research mathematician whose esoteric specialty—nonlinear partial differential equations (which describe such diverse phenomena as the flow of water through pipes and the motion of oil heating in a frying pan)—put him on the frontier of mathematics. In recent years, the focus of his interests has changed to the analysis of practical and theoretical problems in financial economics. He is now a director in the research department of a prominant investment banking firm.

Professor Barbara R. Heller is also an expert in dealing with the problems people have with math, although she focuses on their difficulties in managing money and in making career decisions—practical problems. Trained as an experimental psychologist, she started out working on instrument panel design in single fighter jet planes and the deterioration in reaction time as a function of visual stimuli overload. Professor Heller has devoted the last 35 years to educational challenges. Since 1973, she has served as a senior project director at the Center for Advanced Study in Education. For the past ten years, she has been director of special programs at The Graduate School and University Center of The City University of New York.

Professor Heller now studies the way people learn, be they high school students, college students, or adults. She then designs programs for them, often creating new courses of study and designing new materials that incorporate the new technologies. She is known nationally for her pioneering work in computer-assisted guidance, adult literacy, cooperative education (learning in nonclassroom settings), and teacher training. Recently concentrating on teacher preparation, Professor Heller has conducted major training programs for elementary- through high school-level math teachers, in both public and nonpublic schools and, two years ago, instituted a unique program to train the next generation of college professors who will teach in urban colleges and universities.

Kogelman and Heller have worked together since 1978, mostly on ways to change how mathematics is taught in high school and college classrooms. By training and re-training math teachers, they hope to increase the skills and improve the attitudes of today's students—who will become tomorrow's adults.

The Only Math Book You'll Ever Need is their solution for *today's* adults who want to deal effectively with the math in their lives.

Don't read it from beginning to end. Browse through it. Each topic is self-contained and has all the instructions you need. Read the section you need when you need it. Go through the steps. After a while, you'll see that everyday math really requires mastery of only a few basic ideas. This book reviews these ideas as you need them.

The bulk of the book, which focuses on the things most adults have to deal with in real life, is organized into three major divisions.

Part One is about money and the mathematics of personal finance. The chapter on earnings and taxes covers estimating earnings, finding percent increases, and assessing tax brackets. The banking chapter is comprised of sections on how to balance a checkbook, compute simple and compound interest, and estimate penalties for early withdrawals. Part One also covers investments, such as time deposit certificates, stocks and bonds, and tax-deferred annuities and, in addition, long-term loans, such as mortgages, credit card payments, and installment buying.

Part Two, Outdoor Math, has to do with such topics as eating out, markups and discounts on clothes and appliances, and sales tax. A chapter on foreign travel covers the conversion of currency, temperature, measures (metrics), and electricity. Under hobbies, games and gambling, you'll find out how to adjust the f-stop on your camera and compute odds and probabilities.

Indoor Math situations are covered in Part Three. There's kitchen math, which discusses recipe conversion, timing recipes, and unit pricing. Home improvement shows you how to compute area, buy paint, and order fencing. "Utility" math is the math you need to read your gas or electric meter or phone bill. The section on home computers addresses the question of what they do and who needs one.

We use calculators, and *The Only Math Book You'll Ever Need* shows the keying sequence for many calculations, as well as the paper-and-pencil steps. If you don't already own a small calculator, buy one. Look for a model that has some "memory," a "power" key, and a square root key. Consider buying one that has a "log" button as well. A conveniently sized calculator with all these features sells for between $15 and $25—a worthwhile investment for doing all the math you'll ever need.

PART ONE

THE MATHEMATICS OF PERSONAL FINANCE

1

Earnings and Taxes

Section 1: When Is a Lot a Little?
(Percent Increases/Decreases)

Last week at dinner, our friend announced how pleased he was with his $3,000 raise. The woman sitting next to him said that, in her opinion, $3,000 was not so very special, while Pat described it as "inadequate."

How can the same raise seem so vastly different to these three people?

When you talk about things like raises or other increases in prices, you don't really know how much they represent until you compare them to the *base amount*—that is, to the amount *before* the raise. For example, if the $3,000 increase brought your salary from $27,000 to $30,000, the *percent increase* would be considered significant by most people. However, for someone earning $15,000, an additional $3,000 would be a great boost. But it's only a so-so change for a person with a base salary of $60,000.

Here's how

Evaluating the significance of the raise requires you to compute the percent increase you received. When you ask, "What is the percent increase?" you are really asking, "What percent of the base amount is the increase?"

3

To find the percent increase, first divide the amount of the increase by the old (base) salary and then multiply by 100%:

Percent increase = (amount of increase ÷ base amount) × 100%

In our example, the amount of the raise was $3,000. Let's suppose that the first dinner guest's base salary is $15,000:

Percent increase = (3,000 ÷ 15,000) × 100% = 20%
By calculator:
PRESS 3,000 ⊞ 15,000 ⊟ ⊠ 100 ⊟

In other words, the $3,000 represents a 20% salary increase to this man.

His female dinner companion earns $27,000, so that a $3,000 raise for her means a percent increase of 11.1%:

Percent increase = (3,000 ÷ 27,000) × 100% = 11.1%
By calculator:
PRESS 3,000 ⊞ 27,000 ⊟ ⊠ 100 ⊟

(Incidentally, 11.1% is a good raise, especially in these times, but not nearly as good as 20%.)

In discussing the $3,000 raise, Pat viewed it as fairly paltry in light of the fact that her base salary is $60,000:

Percent increase = (3,000 ÷ 60,000) × 100% = 5.0%

For Pat, the raise would be the equivalent of only a 5% increase over her base salary.

So a raise of the same amount has different meanings depending on how large a portion of the base salary it represents. In our little story, the significance of $3,000 depends on where you sat at dinner—or the base you started from.

EXAMPLE: Now let's consider another example of a percent increase. Suppose your rent went from $650 to $700 a month. The percent increase in your rent is:

Percent increase = (50 ÷ 650) × 100% = 7.7%

The 50 is the difference between $650 (your old rent) and $700 (your new, increased amount).

All percent increase problems are done the same way. Knowing this, let's consider what a 100% increase means. *It is when the amount of the increase is the same as the base amount.* For example, if your rent went up from $650 to $1,300 (which we hope won't happen to you):

Percent increase = $(650 \div 650) \times 100\% = 100\%$

Although it may seem that prices and costs *never* decrease, there are times when this happens and you need to compute the *percent decrease.* *Here's how*

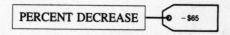

PERCENT DECREASE — -$65

Because you were offered a new job in another company that has better working conditions, higher status and a brighter outlook for advancement, you are willing to consider a salary offer of $32,000, which is less than the $35,000 you are now making. This represents a reduction of $3,000.

To find the percent decrease, divide the amount of the decrease by the base amount (that is, the amount before the decrease), and then multiply by 100%:

Percent decrease = $(3,000 \div 35,000) \times 100\% = 8.6\%$
By calculator:
PRESS 3,000 ÷ 35,000 = × 100 =

EXAMPLE: Let's try another percent decrease problem. As a result of moving to a less expensive location, your rent dropped from $650 to $550 per month.

First, find the amount of the decrease $(650 - 550 = 100)$ and then the percent decrease:

Percent decrease = (amount of decrease ÷ base amount) × 100%
$$= (100 \div 650) \times 100\% = 15.4\%$$

The *process* of finding the percent increase (or decrease) is the same regardless of the context: salary raises, rent reductions, increases in the gross national product, and percent markdowns on merchandise on sale are all instances of this type of problem.

But there's another type of related problem.

Suppose your boss tells you that you are going to be getting a 12% raise but doesn't tell you the *amount* of the raise. If your old salary was, let's say, $25,000, to find the new amount, you must first convert the percent to a decimal.

Here's how

> **CONVERTING PERCENTS TO DECIMALS** ──○ 12% = .12

To convert a percent to a decimal, divide the percent by 100. (That's because the word "percent" means hundredths.) For example:

$$12\% = \frac{12}{100} = 12 \div 100 = 0.12$$

Similarly, 8.25% converts to the decimal 0.0825, like this:

$$8.25\% = \frac{8.25}{100} = 8.25 \div 100 = 0.0825$$

By calculator:
PRESS 8.25 ⊞ 100 ▤
✓*TIP*   You'll notice that dividing by 100 has the effect of moving the decimal point two places to the left, so remember this rule: *to convert a percent to a decimal, move the decimal point two places to the left.*

To figure out the amount of your 12% raise and your new salary, first multiply the base salary by the decimal equivalent of the given percent:

12% of $25,000 = 0.12 × $25,000 = $3,000
By calculator:
PRESS 0.12 ⊠ 25,000 ▤

This gives you $3,000, the dollar amount of the 12% raise. So, the new salary equals $28,000 which is the sum of old salary ($25,000) plus the raise ($3,000).

✓✓*HOT TIP*   You can compute the new salary in *one* operation, like this:

New salary = (% raise ÷ 100) × base salary + base salary
or

New salary = (12 ÷ 100) × \$25,000 + \$25,000 = \$28,000
By calculator:
PRESS 12 ÷ 100 = × 25,000 = + 25,000 =

This one-step calculation gives you the total amount of your salary with the raise, but *not* the amount of the raise.

EXAMPLE: Here's another example to consider. You read that prices have risen 5.5% during the last year because of inflation. This means that goods and services now cost 5.5% more than they did a year ago. If you paid \$350 for a suit last year, what would that same suit cost you today?

First, convert 5.5% to a decimal and calculate the increase:
5.5% = 5.5 ÷ 100 = 0.055
0.055 × \$350 = \$19.25 (the suit will now cost \$19.25 *more*)

Then add the percent increase to the base amount to find the total cost of the suit today:

\$19.25 + \$350 = \$369.25

Remember, you can also compute the cost in one operation.

Total cost = (5.5 ÷ 100) × 350 + 350
By calculator:
PRESS 5.5 ÷ 100 = × 350 = + 350 =

There's another very handy shortcut to solving these percent increase problems, but it requires you to think about the problem somewhat differently.
Here's how

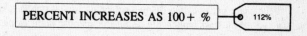

PERCENT INCREASES AS 100 + %   112%

Let's go back to the earlier example of the 12% raise (on a salary of \$25,000) in which we added the amount of the raise (which we had to compute using decimals) to the old salary. In other words, the new salary is equal to the old salary *plus* the amount of the increase.

✓✓*HOT TIP*   The shortcut method is to think of the new salary as being equal to 112%. That is, the value of the old salary (which we will say equals 100%), *plus* 12% (the value of the raise). To find the new salary directly with this method, all that has to be done is to compute 112% of the base salary:

New salary = (100% + % increase) of base salary
= (100% + 12%) of $25,000
= 112% of $25,000   { Notice that the percent still must
= 1.12 × $25,000   { be converted to a decimal.
= $28,000

Now let's do the inflation rate problem by the shortcut method:

EXAMPLE: New cost = [100% (old cost) + 5.5% (inflationary increase)] of base amount
= 105.5% of $350
= 1.055 × $350
= $369.25

It's rather fun to do percent increase problems this way. And it's faster. There's a certain elegance to thinking about a 12% raise as 112% of the base salary, and it's a technique almost guaranteed to impress your dinner companions.

Section 2: How Much Do I Earn?

It is always surprising—and disheartening—to look at your take-home pay after taxes and other deductions are subtracted. Your gross pay—the actual amount you earn *before* taxes and other payroll deductions—appears to be an almost mythical amount of money that never actually passes through your hands. And, indeed, it is mythical for all practical purposes since you can spend only "net" dollars.

When you get over the dismay of seeing how much is taken out of your salary, you may start wondering about the specific deductions and why your paycheck may not be exactly the same each pay period.

Here's how

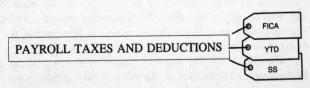

PAYROLL TAXES AND DEDUCTIONS

FICA

YTD

SS

There are at least two, probably three and possibly as many as four different kinds of taxes charged against your salary by government.

First, there is *federal income withholding tax,* probably your single largest deduction. "Withholding" tax, first adopted in 1943, is a method by which the U.S. government collects—in advance—a large proportion of the income tax that it anticipates you will owe based on your salary. The amount of the tax withheld from your paycheck is based on your earnings and the number of dependents claimed on your W-4 withholding form. The more dependents you claim, the less is taken from your salary; the fewer the dependents, the more money is deducted.

While the amount of federal taxes withheld does not vary from pay period to pay period (unless there was a change in your gross salary or in your deductions), the *Social Security tax* deduction can change. Social Security tax, which is different from withholding tax, is also referred to as F.I.C.A. (Federal Insurance Compensation Act). There are currently two components to the F.I.C.A. tax—Social Security (also called Old Age Survivors and Disability Insurance [OASDI]) and Medical ("Medicare Part A"). The medical component was just instituted in 1991 and represents payments toward the Hospital Insurance part of Medicare. The 1992 rate schedule for F.I.C.A. taxes is shown below in Table 1.

Table 1
1992 F.I.C.A. Rate Table

TAX CATEGORY	RATE	MAXIMUM GROSS WAGES	MAXIMUM DEDUCTION
Social Security	6.20%	$55,500.00	$3,441.00
Medical	1.45%	$130,200.00	$1,887.90

As can be seen in Table 1, in 1992 the Social Security tax was calculated at a rate of 6.2% of wages up to a maximum taxable wage of $55,500. By law, this base increases each year to account for inflation. It will reach an estimated $57,900 by 1993 and $67,500 by 1996. Every employee pays the same 6.2% of income up to $55,500. If you earn more than the maximum taxable amount, there is nothing more taken from your salary or credited to your Social Security pension account. The employer deducts Social Security and matches the deduction with a like amount. The self-employed pay a higher rate. In 1992, they paid 12.4% of their income up to $55,500.

The maximum Social Security deduction in 1992 was 6.2% of $55,500 or $3,441. (Since percent means hundredths, 6.2% = 6.2 ÷ 100 = 0.062 and 0.062 × $55,500 = $3,441.00.) If your income exceeded $55,500 and nothing more was deducted, your take-home pay should have increased by the amount of Social Security subtracted from previous paychecks. However, you continue to pay medical tax on income up to $130,200.

Notice in Table 1 that the medical tax is at a lower rate (1.45%) than the Social Security tax. If your gross wages were $55,500 in 1992, the medical tax would have been 1.45% of $55,500, or $804.75. Then, the F.I.C.A. tax you paid was $3,441.00 plus $804.75, or $4,245.75.

If your gross wages were $130,200 or higher, you would have paid $3,441.00 in Social Security tax and $1,887.90 in medical tax (1.45% of $130,200). Thus, the maximum total F.I.C.A. taxes for 1992 was $5,328.90. If you have questions about your Social Security taxes and benefits, call the Social Security Administration's new phone number. It is 1 (800) SSA-1213 (1 (800) 772-1213.)

Since all but nine states currently levy a state income tax or tax on personal earned income (the ones that don't are Alaska, Florida, Nevada, South Dakota, Texas, Washington, Wyoming, New Hampshire, and Tennessee), it is likely that state taxes are also taken out of your gross salary. Like federal income taxes, these state taxes are the same each pay period if there are no changes in claimed dependents or in wage level.

If you live in a city that imposes a city income tax, that too may appear as a regular deduction throughout the year.

Other regular payroll deductions can include the employee's contributions to health, life, and/or disability insurance premiums; savings bonds; credit union dues; union and other agency dues; savings plans and stock plans; certain kinds of loans; and the employee's portion of unemployment insurance.

On an irregular basis, your take-home pay may reflect charges for lost work time (penalties for lateness or absence lower your gross pay and, therefore, your net pay)—just as your gross salary may reflect extra money earned for overtime. Court liens (i.e., salary attachments or garnishments) for outstanding debts can also affect the amount of money you take home.

So, whether your paycheck is generated by computer or handwritten personally by the boss or bookkeeper, because of all the deductions it is a good idea to check them and to do the arithmetic that will tell you whether your *net pay* is correct.

✓*TIP*   To determine your net pay, write down the amount of each deduction (checking each to be sure it is correct and that it applies to you). Next, add up all the deductions to arrive at a total. This total subtracted from your gross pay should equal your net pay for that period.

How do you find out your *gross pay* for a given pay period?

Salaries (which are invariably stated in gross amounts) are usually quoted on an annual, weekly, or hourly basis. Pay periods are usually weekly, biweekly (which means every two weeks, *not* two times each week), or monthly. Many people get paid twice a month. Given your salary for any time period, it is possible to compute your salary for any other time period.

Suppose your salary is quoted on an annual basis and you get paid weekly. You need to find your weekly salary.

Here's how

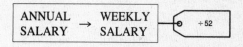

Since there are about 52 weeks in a year (there are actually 52.143 weeks in a year: 365 days $\div 7 = 52.143$), *divide your annual salary by 52 to get your weekly salary.*

EXAMPLE: If your annual salary is $28,500:
$28,500 \div 52 = \$548.08$
By calculator:
PRESS 28,500 $\boxed{\div}$ 52 $\boxed{=}$

If you get paid biweekly, divide your annual salary by 26 (52 weeks $\div 2$):
By calculator:
PRESS 28,500 $\boxed{\div}$ 26 $\boxed{=}$

✓*TIP*   A good *estimate* of a weekly salary is obtained by dividing the annual salary by 50 instead of 52. Dividing by 50 is the type of calculation that can often be done in your head. In our example, the estimated weekly salary on $28,500 would be $570. (Compare this estimate to the actual $548.08 we obtained when we did the calculation based on 52 weeks.)

Now suppose your salary is given on an annual basis and you want to compute your monthly salary.

Here's how

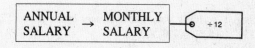

Computing a monthly salary from an annual salary is done by dividing the annual salary by 12 (since there are 12 months in a year).

For example, using our sample salary of $28,500:
$28,500 ÷ 12 = $2,375.00
By calculator:
PRESS 28,500 ÷ 12 =

If you want to compute your hourly pay when your salary is given annually, two steps are required.
Here's how

| ANNUAL SALARY | → | HOURLY SALARY |

To figure out your hourly pay when you know your annual salary, first, convert your annual salary to a weekly salary as we did above, and then divide your weekly salary by the number of hours you work a week. This is your hourly wage.

Step 1 $28,500 ÷ 52 = $548.08 per week
Step 2 (if you work 35 hours a week) $548.08 ÷ 35 = $15.66 per hour
Or (if you work 40 hours a week) $548.08 ÷ 40 = $13.70 per hour

So far we have only done examples where the salary is quoted annually. It works the reverse way too.
If your salary is given on a weekly basis, you can determine how much you'll make in a year.
Here's how

| WEEKLY SALARY | → | ANNUAL | → | MONTHLY SALARY | ×52, ÷12 |

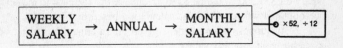

This also is a two-step operation:

Step 1 Multiply your weekly salary by 52 (or by 52.143 to be exact):
$548.08 × 52 = $28,500
By calculator:
PRESS 548.08 × 52 =

Now you've gotten your *yearly* total. Once you know this, you can find your *monthly salary* by dividing by 12:

Step 2 $28,500 ÷ 12 = $2,375.00

Note that although you know the weekly salary, we advise that you do *not* try to convert directly to a monthly salary because there is not an even number of weeks in a month.

✓*TIP* If you feel you *must* convert from a weekly to a monthly salary without stopping to first compute the annual salary, use 4.3 weeks per month to do it (52.143 weeks ÷ 12 = 4.3 weeks):

$548.08 × 4.3 weeks = $2,356.74 = appropriate monthly salary.

Compare this to the actual figure of $2,375.00 we obtained before.

By calculator:
PRESS 548.08 ⊠ 4.3 ⊟

However, if you know your weekly salary, it is easy to find your hourly salary.
Here's how

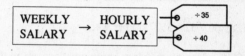

Just divide the weekly figure by the number of hours you work:

$548.08 ÷ 35 or $548.08 ÷ 40 (weekly salary divided by number of hours worked)

If your earnings are computed on an hourly basis, then you can find how much your weekly salary is by multiplying the number of hours you work each week by the hourly amount.
Here's how

HOURLY SALARY → WEEKLY → ANNUAL SALARY ⊸ ×35, ×52

Now let's try a new example with different numbers. Let's say your hourly rate is $8.25 and you work a 40-hour week. Your weekly salary is:

$8.25 \times 40 = \$330$ per week

From this figure you could obtain your annual salary by multiplying by 52 (or 52.143).

Now suppose you know what your salary is on a monthly basis and want to convert this to a weekly salary. You can get an *approximation* by dividing your monthly salary by 4.3 (the average number of weeks per month), but since this is not exact, we suggest you work through an annual salary.

Here's how

MONTHLY SALARY → ANNUAL → WEEKLY SALARY   ⊙ ×12, ÷52

For example, if your monthly salary is $2,200, then your annual salary is:

$2,200 \times 12 = \$26,400$ per year

In this example, your weekly salary is:

$26,400 \div 52 = \$507.69$ per week

✓*TIP*   Sometimes salary approximations are useful. You can quickly estimate an annual salary from a weekly salary by multiplying by 50 (instead of 52). So, if a person earns $100 per week, his annual salary is about $5,000 ($100 × 50). This is a good figure to know because it means that each $100 of weekly salary is the equivalent of about $5,000 of annual salary (and vice versa—each $5,000 of annual salary equals about $100 in weekly salary).

✓✓*HOT TIP*   Knowing this, you can do some quick estimates:

Table 2
Relationship of Weekly and Annual Salaries
(Approximate)

WEEKLY SALARY	⟵——————⟶	ANNUAL SALARY
$ 100		$ 5,000
200		10,000
300		15,000
500		25,000
750		37,500
1,000		50,000

Sooner or later—it usually happens at a particularly boring time at work—almost everyone wonders how much they get paid each *minute*. You can compute this by dividing your hourly salary by 60 (the number of minutes in an hour). Going back to an earlier example:

$15.66 per hour ÷ 60 minutes = $0.26 per minute.

So, if you earn $28,500 a year for a 35-hour work week, your gross salary for each working minute is $0.26! And, if you've spent 45 minutes of company time doing all these computations . . .

Michael Milken, the king of junk bonds who served a jail sentence, was the highest paid financier in history. In 1987 alone, he made $550 million. That comes to $132,211 per hour (assuming an 80-hour week), which works out to $2,204 *per minute*. If it's any consolation, his per minute pay as a typical federal prisoner was less than $0.01.

(Incidentally, many things are figured on the basis of minutes or even seconds. For example, each 30-second advertising spot during the telecast of Super Bowl XXVI on January 26, 1992 was sold for a record $850,000. That's $28,333 *per second*!)

Section 3: More = More (The Meaning of Tax Brackets)

"Our constitution is in actual operation, everything appears to promise that it will last; but in this world nothing can be certain, except death and taxes."

While there is the truth of the inevitability of taxes, as Benjamin Franklin wrote to a friend in 1789, there are many untruths about them as well.

The untruth that we will explore here is one heard frequently: "It doesn't pay for me to earn more [or, alternatively, for my spouse to work] since that would put me into a higher tax bracket where I'd make less than I did before." This myth is the result of a misinterpretation of tax brackets.

Earning more money may indeed place you in a higher tax bracket, but only with respect to *some part of the additional income*. You may pay a higher rate on the *last dollars* earned, but because of the way the tax rate schedule is graduated, the money you made before the increase is not taxed at the higher rate. Making more money *always* means having more available after income taxes—there is no way you can end up with less.

Here's how

TAXABLE INCOME

In any discussion of income taxes, the first thing you need to remember is that taxes are not paid on your total gross income. There are adjustments to income and allowable deductions that can and should be subtracted from the total money you earn.

Allowable *adjustments* to income may include such things as alimony payments and payments into a Keogh, SEP (Simplified Employee Pension Plan), or IRA plan (Individual Retirement Account—we'll talk more about these tax-deferred savings plans in Chapter 3) and other items such as self-employed health insurance payments and self-employment tax listed in the Internal Revenue Service (IRS) Form 1040—the booklet the IRS sends taxpayers each year. There adjustments to income result in your modified adjusted gross income (AGI).

Allowable *deductions* include subtractions from your income for dependents, for certain charitable contributions and medical expenses, and for many other items. Your accountant or the free IRS consulting service offered in many cities (or by phone any place) can supply you with complete guidelines about income adjustments and deductions. You should also take advantage of the many books and other free literature that explain allowable deductions. One of the more helpful is the free IRS publication #17, "Your Federal Income Tax," which is available from the Forms Distribution Center in your state. Consult Form 1040 for the address.

✓*TIP*   Be sure to get up-to-date editions of tax publications because tax laws change regularly and what may have been a legitimate item one year may be disallowed the next. (The reverse is also true: new allowable deductions may be added.)

The amount of income remaining after adjustments and deductions have been taken is called *taxable income*. The tax bracket you are in (and, therefore, the taxes you pay) is based on this figure. Increasing your deductions decreases your taxable income so it makes sense to carefully explore all possible deductions for which you are eligible.

Now that we have defined taxable income, we're ready to examine tax brackets—how they are constructed and what they mean. Let's start with tax rate schedules.

Here's how

TAX SCHEDULES

Included in the IRS Form 1040 booklet are several *United States Federal Income Tax Schedules:* one for single taxpayers, unmarried heads of households, married taxpayers filing joint returns and qualifying widows and widowers, and married taxpayers filing separate returns. These different tax schedules (another name for charts or tables) apply to different categories of people who are taxed at somewhat different rates for the same income. The first thing to do is to find the rate schedule appropriate for you.

Suppose you are a married taxpayer filing jointly and the combined taxable income for you and your spouse is $55,000. You would use Tax Schedule Y–1 (reprinted in Table 3 below) to compute the amount of your tax.

To compute the taxes on the combined income of $55,000, look down column (a), the "Taxable income over—" column, until you find the dollar amount *closest to but below* your taxable income. In our example, that would be $34,000—closest to but less than the $55,000 we are using as an illustration.

Next, read across the row of the table to the second column (b), "But not over—." In our example that amount would be $82,150. This number serves as a check for you because your $55,000 taxable income actually does fall over $34,000 and is not over $82,150.

Now go on to the third column (c) and fourth column (d), which together tell you exactly how to compute your tax. In our case, the third and fourth columns say, "Compute $5,100.00 + 28% of the amount over $34,000." To figure out "the amount over" $34,000, take your income and subtract $34,000:

$$\text{Your income} - 34,000 = \$55,000 - \$34,000$$
$$= \$21,000 \text{ (or the amount you'll pay 28\% of)}.$$

Now, your total tax according to the tax rate schedule direction is computed as follows:

$$\begin{aligned}
\text{Tax} &= \$5,100 + 28\% \text{ of } \$21,000 \\
&= \$5,100 + (0.28 \times \$21,000) \\
&= \$5,100 + \$5,880.00 \\
&= \$10,980
\end{aligned}$$

Table 3
1991 United States Federal Income Tax Rate Schedule Y–1
(for "Married Taxpayers Filing Jointly and Qualifying Widows and Widowers")

(A) TAXABLE INCOME OVER—	(B) BUT NOT OVER—	(C) COMPUTE	(D) OF THE AMOUNT OVER—
$ 0	$34,000	15%	$ 0
34,000	$82,150	$ 5,100 + 28%	34,000
82,150	—	18,582 + 31%	82,150

So the federal tax on $55,000 is $10,980.

> *By calculator:*
> PRESS .28 ⊠ 21,000 ▣ ⊞ 5,100

In this calculation, we first computed 28% of $21,000, then we added $5,100.00. Note that 28% is the *highest* percent or highest tax rate you pay on an income of $55,000, but you don't pay it on the total taxable income. The highest rate or percentage is what is meant by your *federal tax bracket*. You moved into the 28% bracket when your taxable income fell between $34,000 and $82,150. Had you earned just a little less than $34,000, the highest tax rate would have been 15%—that is the percentage in the row above in the rate schedule, Table 3.

Going back to the example, even though you understand what the 28% means, you may ask, where did the $5,100 come from? To find out, we need to go back one row in Table 3. Notice that for income below $34,000, the taxes are computed as 15% of the amount over $0. If your taxable income was $34,000, you would have the highest permissible amount over $0. On this amount, you compute 15%:

15% of $34,000 = 0.15 × $34,000 = $5,100

Thus, the *maximum amount* of tax to be paid on incomes up to $34,000 is precisely the amount in column (c) for the income range $34,000 to $82,150.

Starting at the top of Table 3, you will see that if your taxable income is below $34,000, the tax rate (for 1991) was 15%. For the next earnings range, the tax was $5,100 (the taxes you'd pay on an income of $34,000) *plus* 28% of the last dollars earned. Thus, the dollar amounts in column (c) are the total taxes on the income up to the amount in column (a), and the percentages—28% and 31%—apply to the part of your income which is over the column (a) amount but not over the column (b) amount.

This is what a *graduated tax structure* means: each portion of your taxable income is taxed at a different and increasing amount up to a taxable income of $162,770. At this level, and thereafter, some special computations must be done to determine the tax on the amount over $162,770. In other words, regardless of your total taxable income, everyone pays 15% of the first $34,000, 28% of the amount between $34,000 and $82,150, and 31% of the amount over $82,150. The highest rate that you use is referred to as your federal income tax bracket.

Now, let's examine what a raise to a new, higher income means.
Here's how

MOVING TO A HIGHER BRACKET

In our earlier example, you were in the 28% bracket with a taxable income of $55,000. Now, let's suppose you received an important promotion and increased your income by $30,000 with no increase in deductions. This would give you a total taxable income of $85,000.

Turning back to Table 3, we find the row in which this income falls: "over $82,150." In columns (c) and (d) we see that the tax is:

$18,582 + 31% of the amount over $82,150.

Since $85,000 is $2,850 more than $82,150, we must compute 31% of $2,850:

$$Tax = \$18,582.00 + 31\% \text{ of } \$2,850$$
$$= \$18,582.00 + 0.31 \times 2,850$$
$$= \$19,465.50$$

So, the total federal tax on an income of $85,000 is $19,465.50

By calculator:
PRESS 0.31 ⊠ 2,850 ⊟ ⊞ 18,582 ⊟

Again notice that the highest tax rate (31%) only applies to the last $2,850 in income, even though it was a raise of $30,000 that put you into this bracket. Thus, part of the "last dollars" you earned moved you into a new tax bracket, but the new tax rate only affects the tax you pay on the portion of *that* money and not on the money you were earning before the raise.

So, when people say that having a spouse work "isn't worth it because the additional earnings mean higher taxes," they are correct only to the extent that a higher rate is imposed on the *extra* income. What they probably mean is that what's left of this additional income after taxes have been paid does not cover the expenses incurred in going to work—travel costs, housekeeper or baby sitter, new clothes, and so on. In our example, a $30,000 increase in income resulted in a tax increase of $8,485.50 ($19,465.50 − $10,980.00). This means you got to keep $21,514.50 ($30,000.00 − $8,485.50). This is still a very substantial increase.

There are tax tables in which the exact amount of tax is calculated for all taxable income up to $50,000, assuming a standard amount of deductions. But since the arithmetic has already been done for you, you can't determine your tax bracket from these tables.

The federal tax bracket you're in doesn't tell the whole tax story. There may be state and city income taxes as well. State and city taxable income is usually somewhat different (but not dramatically so) from your federal taxable income because of different allowable adjustments and deductions. The tax rates are built on the same kind of graduated scale, however, but state and city rates are generally quite a bit lower than federal rates. For example, for 1991 the highest federal rate was 31%, while the highest New York State income tax rate was 7.875%, and the highest New York City income tax rate was 4.46%.

Since these local income taxes can add as much as 12.3% to your tax bill, your overall rate of taxes should take all taxes into account. If you lived in New York City and were in the highest tax brackets in 1991, you were paying about 43.3% (31% Federal + 7.875% State + 4.46% City) of your "last dollar in income" in taxes.

That's not the worst of it. Earlier in this chapter, we pointed out that Social Security tax is computed at the rate of 6.2% of your gross income up to $55,500 and F.I.C.A. medical tax is computed at 1.45% of your gross pay up to $130,200 (for 1991). These taxes are, of course, in addition to federal, state, and local income taxes.

In our hypothetical examples you can easily end up paying about 50% of your last dollar in taxes.[1] This means you keep only about 50¢ out of a dollar—out of your last dollar. Remember, not all of your income is taxed at this rate: your first dollars of income are taxed at lower rates and those rates are unaffected by any increases in income. So, the truth is, with additional income you always have more money available—but not as much more as you would like!

---

[1] The 50% total tax bill is only an approximation. The computation of the actual tax on the last dollars of income is a bit more complicated. Within certain limitations, state and local taxes are deducted from the federal tax, thereby lowering the federal tax bill slightly. In addition, the F.I.C.A. medical tax continues (on increases up to $130,200), while Social Security tax stops when income exceeds $55,500.

2

Banking

Section 1: Checking It Out: How to Balance Your Checkbook

There are a good many people who, like our friend Arlene, don't ever balance their checkbooks. Many don't even keep records of the checks they write. Why should they bother with these recordkeeping tasks? After all, as Arlene points out, "the bank sends you a statement each month and returns your checks," and, in most cities, automatic teller machines (ATM's) allow you to just insert your bank card for a read-out of your balance.

There are three reasons for *carefully* recording the checks you write and *regularly* balancing your checkbook:

1. What you write in your check register (or on your check stubs) is a permanent record of your transactions. This record is absolutely essential if you want to "stop payment" on a check.
2. Banks can and do make mistakes.
3. Your running "balance forward" is the only accurate, up-to-date estimate of how much you really have in your account at a given time. The automatic teller machines tell you only what has cleared the bank.

Here's how

THE CHECKBOOK REGISTER

Every checkbook, whether for personal or business use, has space for you to write down important information pertaining to each check. Such information usually includes the number of the check, the amount, the date, the name of the person or business to whom the check was drawn, and the "purpose" of the check. A brief description of the purpose of the check is particularly helpful in preparing tax returns since the expense may be deductible.

A typical checkbook register looks like this:

Number	Date	Transaction	Fee	Add (+)	Subtract (−)	Balance Forward
__ __	__ __	To: _____	__	__	__	__
		For: _____				
__ __	__ __	To: _____	__	__	__	__
		For: _____				
__ __	__ __	To: _____	__	__	__	__
		For: _____				

Since it only takes a few moments to fill in the register *immediately after writing a check,* why don't all people do it?

All too many people have gotten into the bad habit of carrying loose checks around with them. They don't carry their personal checkbook and register or business account checkbook and register with them because they're too big or inconvenient, but they do carry loose checks "just in case." Also, joint account checkbooks are typically carried by one partner or spouse, or, more likely, left at home or in the office.

There are simple solutions to these problems. If your checkbook is inconvenient to carry for one reason or another, ask your bank for a smaller one that you can keep with you. Then you can fill in the register as you go along. (But you still must remember to transfer these records to the primary checkbook register.) If you have a joint account, ask the bank for an extra checkbook and register so that both parties can carry a checkbook and record all checks as they're written.

The checkbook register enables you to record the fee (if any) for each check you write, as well as cash withdrawals (which can be considered as checks to yourself for cash and should be handled as if they were such checks), deposits, and errors. We'll come back to errors later.

Suppose your previous balance was $172.20 when you wrote a $15 check (on January 2, check #150) to the motor vehicle bureau and then paid $46.75 (on January 2, check #151) to your hairdresser. The next day you wrote a check to the electric company for $48.52 (on January 3, check #152) and made a deposit of $582.50. If your bank charges a fee of $0.10 per check, your register would look like this:

Number	Date	Transaction	Fee	Add (+)	Subtract (−)	Balance Forward
		To: For:				172.20
150	1/2	To: Department of Motor vehicles For: Car Registration	.10		15.00	15.00 157.20
151	1/2	To: Cutting Impressions For: Haircut	.10		46.75	46.75 110.45
152	1/3	To: Con Edison For: Electric Bill	.10		48.52	48.52 61.93
	1/3	To: Deposit For:		582.50		582.50 644.93
	1/4	To: ATM For: Cash - Self	.25		60.00	60.00 584.43

So far, so good?

If you've kept this kind of record, you'll have all this information (not proof—the cancelled checks are proof) to help you prepare your tax returns. And, by filling in the "Balance Forward" column, you'll always have an exact record of how much money you have in your checking account at any time.

Now we'll explain how we have filled in the register so that all we have to do to compute how much we actually have in our bank account at any time is to add or subtract accurately.

Here's how

"BALANCE FORWARD"

From the previous pages' transactions, we started with a balance forward of $172.20. We then made two deductions, one for $15.00 and one for $46.75. By subtraction, the

Balance forward = $172.20 − $15.00 = $157.20
Balance forward = $157.20 − $46.75 = $110.45

After paying the electric bill, we had a running

Balance forward = $110.45 − $48.52 = $61.93

Next, we add in our deposit,

Balance forward = $61.93 + $582.50 = $644.43

so that on January 3 we have a total of $644.43 in our hypothetical account. But that evening we noticed we were almost out of cash, so we withdrew $60.00 from the ATM at the corner bank. We choose to treat this transaction as if we wrote a check to ourselves for cash (which, in fact, *is* what an ATM withdrawal is), and we enter it in our checkbook register that way, remembering also to note the $0.25 fee.

✓*TIP*   Notice something important! While we have recorded the check and ATM fees ($0.10 per check, $0.25 for using the automated teller machine), *we have not made them part of the running balance*. When we balance the checkbook, we'll account for these fees, but, on a daily basis, it's too confusing to do so and too easy to forget to subtract them. However, for accuracy, our account really has $0.55 less in it (3 checks × $0.10 per check, plus $0.25 for the ATM fee) than the balance forward indicates.

Keeping your checkbook register filled out systematically is a necessary step in reconciling your accounting with the bank's accounting.

There are several reasons why you and the bank may not agree on how much you have in your account. Sometimes it's because either you or the bank (or both) made an error. You might have recorded the amount of a check or deposit incorrectly, or you might have subtracted or added incorrectly. The bank can and also does make recording errors, but you can be certain the bank did not make an arithmetic error: the operations of addition and subtraction are computerized and computers are perfect at these routine computations—provided they are fed the correct numbers.

The most common reason for a discrepancy between your balance and the bank's statement of your balance has to do with what checks and/or deposits have *cleared* the bank. Your balance forward includes *all* checks, cash withdrawals, and deposits. The bank statement may not. If a person or company is slow in cashing your check, for example, the bank may not have subtracted that amount from your account and will show a bigger balance than you actually have. If a deposit made just before the statement was issued has not yet been recorded by the bank, the statement will show your account balance as less than it actually is.

The process of reconciling your checkbook with the bank statement involves accounting for bank fees, missing checks, and unrecorded deposits. It also requires finding mistakes in recordkeeping and/or computation.

Here's how

BALANCING YOUR CHECKBOOK

Every month, the bank sends you a statement to tell you what you have in your account according to its records. (ATM statements can do this on a more frequent basis.) All the checks you have written that have cleared in the last month are also returned to you.

Balancing your checkbook takes several steps, and we suggest doing them in order. Many bank statements have a form on the back to help you balance your checkbook. The first few times, follow these 10 steps as you would a recipe.

Let's start.

Step 1 *Deposits:* Go through your register and check off all deposits you made (whether in person, by mail, or through an ATM) that appear on the bank's statement. After you've done this, you will have isolated any deposits you made that the bank has not yet credited to your account. (These will be the unchecked deposits in your register.)

✓*TIP*   Generally, banks handle deposits promptly, but it is still possible that a recent deposit won't show up on this month's statement. If it doesn't appear on the following month's statement, contact your bank immediately. It may have been misplaced.

As you're comparing deposits, make certain that the amount of each one as recorded by you and by the bank agree. If they don't, go back to your original deposit slip. If the mistake was the bank's, let them know as soon as possible. (We'll tell you what to do about a mistake of yours later, under "Finding Errors.")

Step 2 *Canceled checks:* Start by arranging the returned checks in numerical order (the order you wrote them in).

Go through your check register and check off each cleared check. As you do so, make sure that the amount of the check listed by the bank corresponds to the amount you recorded.

Compare the amount of each cleared check with the amount the statement lists. In this way, you should be able to locate any *transpo-*

sition errors, the most common type of error you or the bank is likely to make. (A transposition error consists of reversing two numbers, like this:

actual amount   1⑦9⑥.54
transposed figure   1⑨7⑥.54

Transposition errors are *very* hard to spot.)

At the end of Step 2, your register will have unchecked items—deposits that have not yet been credited to your account and checks that have not yet been debited to your account (called "outstanding" checks).

Step 3 *If you are charged a fee for each check you write* (otherwise, go right on to Step 4): Count the number of checks that were returned to you and multiply this number by the fee you are charged for each check ($0.10 in our example). Then count the number of withdrawals you made from ATM's and add the fees for this service. The totals will have to be deducted from your balance. More about this later.

Step 4 *Cash withdrawals:* If, as we recommend, you have been considering cash withdrawals as a form of check to yourself, they should be recorded in your register.

Compare your bank statement and register, checking off cash withdrawals that appear on the statement. If you find one *you* forgot to record, you'll have to adjust for it (subtract it) later; if you find one the *bank* hasn't cleared, you'll have to add it in when balancing your account.

At this point, all the items that appear in the statement will have been checked in your register—except for fees. You probably have some outstanding checks and maybe an uncredited deposit. If you've been careful, you may have caught a recording error.

Depending on the kind of checking account you have, you may be charged a regular monthly fee, or there may be a charge for new checks, or deposit slips, or a penalty for returned checks that did not clear for one reason or another. These fees, like per check and ATM charges, appear on your statement and will also have to be deducted in the final steps of balancing your checkbook.

Step 5 *The starting point:* Crucial to balancing your account is starting at the right place. *Start at the last item in your register that you checked.* Write down the balance that remains after the last checked item. Let's suppose that that balance is $929.71 (after check #160).

Step 6 *Outstanding checks:* List and total all the checks that are outstanding (they should be the items with *no* check mark that come before the last checked item in your register). For example, let's suppose your outstanding checks add up to $251.55.

Number	Date	Transaction	Fee	Add (+)	Subtract (−)	Balance Forward
		To:				
		For:				996.09
160	2/1	To: NY Telephone	.10		56.38	56.38
		For: Phone Bill ✓		(last checked item)		929.71
161	2/3	To: Sear's			216.47	216.47
		For: Lamps				713.24
		To:				
		For:				

Step 7 *Cash withdrawals:* Now do the same thing for cash withdrawals *you forgot* to enter in your register. Suppose these total $120.

If you have a record of a cash withdrawal that has not appeared on your statement, you must note it and add it to the sum of the outstanding checks that was computed in Step 6.

Step 8 *Uncredited deposits:* Make a list of any deposits that were not checked off. Find the total. For our example, let's suppose this comes to $92.50.

Step 9 *Reconciling the balance:* First, add the sum of the outstanding checks (Step 6) to your starting balance (Step 5):

$$\$251.55 + \$929.71 = \$1,181.26$$

The reason for this is that you subtracted this amount from your register balance, but the bank did not clear these checks. By adding, you are adjusting your total to better accord with the bank's total.

Then subtract the sum of the "forgotten" cash withdrawals (Step 7) from your balance. Again, you do this to bring your total into closer accord with the bank's total:

$$\$1,181.26 - \$120.00 = \$1,061.26$$

Next, the sum of the deposits that were not checked off (Step 8) must be subtracted from the balance because you added this to your register, but the bank has not yet done so. By subtracting this amount, you are bringing your total even closer to the bank's:

$$\$1,061.26 - \$92.50 = \$968.76$$

The amount of $968.76 is still not the amount that the bank shows as your closing balance this month because of check fees and service charges. Suppose these total $3.55. Subtract $3.55 as follows:

$$\$968.76 - \$3.55 = \$965.21$$

If all has gone well, this is precisely the amount that appears as the "closing balance" on your statement. If it is, you know that you and the bank agree. The account is balanced. However, in our case, there is more to do.

Step 10 *Odds and ends:* Notice that while you considered the service charge in reconciling the account, you never actually wrote this charge down in the register as part of the running total. It's necessary to do so now so it need not be considered again when you get your next statement. Use the first empty line. Identify the period covered and subtract the amount of the service charge from your balance forward.

Also, it's good practice to make a note in the register at the point at which the account balanced—the point of the last checked item. In our example this was at $929.71.

Go back in the register and put a circle around this item to identify it as the balancing point and write something like, "OK to here." When your next bank statement comes, you'll know that any error made occurred after this point.

Arlene says this is all well and good, but what if the account does not balance after all this work? In that case, you must try to find the source of the error.

Here's how

FINDING ERRORS

If it is a bank error, inevitably, it was in *crediting or debitting* your account by an incorrect amount. If you've worked through Steps 1–4 comparing your records and the bank's, you will have located the bank's mistake. Notify them at once; some banks give you as many as 30 days to report errors, but for some, the limit is 14 days (except for electronic transfers).

If the error is yours, it is most likely to be a *transposition error* that occurred when you transferred a number from one page in your register to the next. *Look for these errors first.*

The next most common type of error is *adding* a check when you meant to subtract it—or, similarly, *subtracting* a deposit that should have been added. Look carefully to be sure you added when you meant to and subtracted when that was the appropriate operation.

Arithmetic mistakes in adding or subtracting are not as frequent as you might suspect. They happen less often than the other kinds of mistakes. But if you still haven't found an error in copying or in crediting or debitting, you need to recheck all of your additions and subtractions.

Finally, an error can occur because you did not account for *all outstanding checks or deposits*. You may not have gone back far enough in your checkbook register to pick up an item from several months ago, such as a check that has not yet cleared. If the item concerns a deposit, get in touch with your bank. If it's a check, you might want to contact the person or business you wrote it to to see if it was received or whether you have to stop payment and issue a replacement.

Let's assume you found an error. There are two ways to correct it. The long way involves going back and starting at the point the error was made. If it was in subtraction, you will have to make the correction in each subsequent balance. Then, of course, you will have to reconcile the account again, starting at Step 5!

A less time-consuming way to make a correction is to figure out how much you were off by—too much or too little—and add or subtract that amount using the next blank line in your register. Suppose your mistake was originally in adding a check for $100, rather than subtracting it. The result is that you gave yourself credit for $200 too much ($100 when you added it incorrectly and $100 for not subtracting it correctly). So, to compensate for this mistake, you would have to subtract $200 from your register.

In our example we have to make two corrections. The first is because we forgot to record $120 of cash withdrawals. We also forgot to subtract $3.55 in check fees.

To illustrate the first correction, if the last check you wrote was to Joe's Auto Body shop for $190.00, leaving a balance in your account of $434.19, the error would be made on the next line in your check register as follows:

Number	Date	Transaction	Fee	Add (+)	Subtract (−)	Balance Forward
		To:				624.19
		For:				
168	2/12	To: Joe's Auto Body	.10		190.00	190.00
		For: Car Repairs				434.19
		To:			120.00	120.00
		For: Error - Forgotten Cash Withdrawal				314.19
		To:				3.55
		For: Check Fees				310.64

Arlene's checkbook is in terrible shape—she hasn't balanced it since she opened it last year. To balance it now is a horrendous, tedious task and probably an impossible one. She might as well accept the correctness of one more statement from the bank—which by now may include several errors for or against her (the ones *for* her she could probably live with)—before she resolves to take charge of her affairs.

So do what Arlene does. Forget the past and move ahead. Open a new checking account in the same bank or another one. As long as you're doing this, you might want to do some research and find a more convenient bank—located where you need it—or one with lots of cash machines or extra-long hours. You might also seek a bank that has lower (or better yet, no) monthly service charges, low per-check check charges, or no required minimum balance. Look for a bank that pays you interest on your balance. Receiving no interest on a required minimum balance is tantamount to paying a fee since *the bank* is earning interest on your balance, interest that you could be earning yourself.

You can open the new account immediately, but don't close out the old one until all of the checks have cleared. Then you'll probably have to write one last check to yourself for the final balance, subtracting the fees that you will owe. If you're unsure, bank officers are there to help.

Section 2: A Matter of Interest (Simple and Compound Interest)

When you plan to borrow or to invest money, you need to be concerned with interest. Interest rates matter because interest is a *fee*—either charged to you or earned by you as the case may be.

You are investing money with a bank and, in effect, lending the bank or other financial institution money and earning interest when you:

- Deposit money in a savings account
- Buy a certificate of deposit (CD) or savings certificate
- Invest in a money market fund
- Defer your income through an IRA or Keogh plan

The interest you earn is the bank's payment to you for the use of your money. It is to your advantage to get the *highest* rate of return on the money you "loan" to the bank.

The bank, in turn, uses your money by investing it in loans to other people and businesses. When money is *borrowed* from the bank, the borrower is charged a fee—interest—for the use of the money. As a borrower, you will be looking for the *lowest* interest charges.

- A mortgage, for example, is a bank loan to the potential purchaser of a house.
- A car loan (another type of mortgage) is a loan for the purchase of an automobile.
- A student loan enables the borrower to "buy" college tuition.

The mortgagee, car buyer, and student each pay the bank interest for the use of money in much the same way that the bank pays *you* interest for the use of *your* money. The bank makes a profit by charging more interest on the money it loans than it pays on the money it borrows.

There are two kinds of interest: simple and compound. *Simple interest* is relatively easy to understand and compute, but it is rarely used. However, in order to understand *compound interest,* it makes sense (no pun intended) to start with an explanation of how simple interest works.

Here's how

SIMPLE INTEREST

Simple interest is usually quoted on a yearly basis: "9% simple interest" means that you would earn 9% of your investment in interest *per year*. Let's suppose you decided to invest $500 at this rate. Your interest, after a year, would be:

$$I \text{ (for interest)} = 9\% \text{ of } \$500 \text{ for 1 year} \quad \left\{ \begin{array}{l} \text{percent means hundredths,} \\ \text{so } 9\% = 9 \div 100 = 0.09 \end{array} \right.$$
$$= 0.09 \times \$500 \times 1$$
$$= \$45$$

There is a formula for simple interest:

$$I = Prt$$

The Prt means $P \times r \times t$ (mathematicians are fond of leaving out the multiplication sign), where:

I means *Interest*
P is the *Principal,* or amount of money with which you started

r is the annual *rate* of interest
t stands for the *time* the money is invested at the rate of interest (r)

Let's apply the formula to the first example again, but now we'll only deposit the money for 6 months (which is $\frac{6}{12} = \frac{1}{2}$ of a year, or 0.50). We would then earn:

I = $500 × 0.09 × 0.50
  = $22.50

✓*TIP*   A shorthand way of looking at this problem is to consider that in 6 months you would earn one-half of 9% or 4.5% of your $500 in interest:

$$I = 4.5\% \text{ of } \$500 \atop = 0.045 \times \$500 \left\{ \; 4.5\% = \frac{4.5}{100} = 0.045 \text{ (because \% means hundredths)} \right.$$
$$= \$22.50$$

The next example of simple interest involves investing $1,200 at 8% for one year and 3 months. How much interest (I) would you earn? Substituting in the formula, we have:

$$I = 1,200 \times 0.08 \times 1.25 \left\{ \begin{array}{l} \text{Rate (r)} = 8\% = \dfrac{8}{100} = 0.08 \\ \text{Time (t)} = 1 \text{ year and 3 months} \end{array} \right.$$
$$= 120 \qquad\qquad\qquad = 1\frac{1}{4} \text{ year} = 1.25$$

So you would earn $120 interest on $1,200 invested at 8% for one year and 3 months.

By calculator:
PRESS 1200 ⊠ .08 ⊠ 1.25 ▣

That's all there is to simple interest. But interest rates are generally compounded. *The difference between simple interest and compound interest is that with compound interest you earn interest on your interest.*
Here's how

COMPOUND INTEREST

Suppose a bank offers interest at a quoted rate of "12% compounded monthly." This means you earn one-twelfth of 12% or 1%, each month.

Let's start with a deposit of $500 that you leave in an account for 6 months. Each month, 1% interest is credited to your account. In each successive month, you earn 1% of the amount you have in your account at the end of the previous month. This amount will include interest that has accumulated each month. The table below (Table 1, page 33) shows the interest you'd earn for each of the 6 months.

Notice that each month the amount of interest increases—$5.00 for the first month, $5.05 for the second month, and so on. This is because the balance (or principal) is increased each month as a result of the addition of earned interest. In the first month, 1% interest is earned on $500, but in the second month, the interest is computed on $505. Continuing this way, at the end of 6 months, you have $530.76 in your account.

Compare this now with how much simple interest is earned on $500 for 6 months:

$$I = Prt$$

$$I = 500 \times 0.12 \times 0.50 \ \{12\% = \frac{12}{100} = 0.12; \text{ and}$$

$$6 \text{ months} = \frac{6}{12} \text{ years} = 0.50 \text{ years}$$

$$I = 30$$

So, with simple interest, you would earn $30, whereas with compounding, the total interest after 6 months is $30.76. ($530.76 − $500.00). The additional $0.76, the result of compounding, is the interest earned on the interest.

Big deal—$0.76! Truly a small difference. But that's only because of the small amount of principal we started with and the short period of time we invested it. All other things being equal, the longer the time period, the

Table 1
Compound Interest (1% Monthly)

END OF MONTH:	OLD BALANCE	INTEREST (1% OF OLD BALANCE)	NEW BALANCE (OLD BALANCE + INTEREST)
1	500.00	$0.01 \times 500.00 = 5.00$	505.00
2	505.00	$0.01 \times 505.00 = 5.05$	510.05
3	510.05	$0.01 \times 510.05 = 5.10$	515.15
4	515.15	$0.01 \times 515.15 = 5.15$	520.30
5	520.30	$0.01 \times 520.30 = 5.20$	525.50
6	525.50	$0.01 \times 525.50 = 5.26$	530.76

greater the difference between the interest earned at a simple rate and that earned at a compound rate.

The computation of compound interest in the way we showed it in Table 1 (by doing 6 simple interest calculations) becomes more time-consuming as the number of compounding periods increases. Imagine how many calculations would have to be done if your deposit was compounded *daily* for the 6 months of your deposit!

There is a formula that allows us to calculate in one operation what the final balance would be ($530.76). The *compound interest formula* is:

$$S = P(1 + i)^n$$

It is read, "S equals P multiplied by 1 plus i to the nth power." Here's how the formula works:

S = the final amount of money (principal *plus* interest)
P = the *Principal,* or the amount we started with (in our example, P = $500)
i = *the periodic interest rate divided by 100.* The periodic interest rate is the quoted interest rate divided by the number of times per year that the interest is compounded. To get i, the periodic interest rate is divided by 100. (In our example, the periodic interest rate is 12% ÷ 12 (months) = 1%, and i = 1 ÷ 100 = 0.01).
n = *the number of interest periods* (6 in the example because the money was deposited for 6 months and interest was credited monthly)

Let's substitute our numbers in the formula:

$$S = 500 \ (1 + 0.01)^6$$

First note that $(1 + 0.01) = (1.01)$ and that $500 \ (1.01)^6$ means $500 \times (1.01)^6$.

Raising a number to a power (6 in this case) means multiplying that number by itself *that* many ("6") times, so:

$$(1.01)^6 = 1.01 \times 1.01 \times 1.01 \times 1.01 \times 1.01 \times 1.01$$

You can do this calculation by hand or by using your calculator:

PRESS 1.01 ⊠ 1.01 ⊠ 1.01 ⊠ 1.01 ⊠ 1.01 ⊠ 1.01 ⊟

but you can easily lose count—especially as "n" gets larger (say, if the money was left on deposit for a year and n = 12, or if interest is compounded *daily* for a year and "n" becomes 365).

There is a type of calculator that does this computation faster and in one step. It's the kind that has the capacity to raise any positive number to a *power*, and it's recognizable because it has a button marked either:

$\boxed{Y^x}$ or $\boxed{X^y}$

the positive number Y is to be   the positive number X is to be
raised to the "xth" power   raised to the "yth" power

This type of calculator is not very expensive and allows you to compute $(1.01)^6$ as follows:

PRESS 1.01 $\boxed{Y^x}$ 6 $\boxed{=}$ and the result 1.06152 appears.

So, to finish substituting in the formula:

$S = 500 \times 1.06152 = 530.76$

This amount of money (principal plus interest) is our balance at the end of 6 months.

EXAMPLE: Let's do a new example where we don't know the answer beforehand. You have $2,000 in an account that pays interest at the rate of 10% compounded quarterly. If you leave the money on deposit for one year, how much will you have in your account?

SOLUTION: To answer this question, we apply the compound interest formula. The first step is to identify the quantities P, i, and n. In this instance:

$P = 2,000$

i is obtained by first computing the periodic interest rate, which is $10\% \div 4 = 2.5\%$ (since interest is compounded quarterly, there are 4 compounding periods per year), and then dividing 2.5 by 100. So $i = 2.5 \div 100 = 0.025$

$n = 4$, since the deposit remains in the bank for 4 interest periods

Substituting, we have:

$S = P (1 + i)^n$
$S = 2,000 \times (1 + 0.025)^4$
$S = 2,000 \times (1.025)^4$

Next, 1.025 must be raised to the 4th power:

By calculator:
PRESS 1.025 $\boxed{Y^x}$ 4 $\boxed{=}$ The result is that $(1.025)^4 = 1.1038$

Continue substituting in the formula:

$$S = 2,000 \times 1.1038 = 2,207.60$$

At the end of the year, there will be $2,207.60 in your account, which means that you earned $207.60 in interest ($2,207.60 − $2,000.00) for the year. Compare this to 10% simple interest:

$$I = Prt$$
$$= 2,000 \times 0.10 \times 1 = 200$$

Thus, in one year, the effect of compounding is an extra $7.60 in interest.

A compound interest rate always produces more money than the same simple interest rate. All other things being equal, the more frequent the compounding, the more interest you earn. Let's test this by comparing the interest earned in one year on $100 at 10% compounded quarterly, with the interest earned on $100 at 10% compounded monthly, and at 10% compounded daily. In the first case, compounded *quarterly,* we have:

$$S = 100 \times (1 + 0.025)^4 = 100 \times 1.1038 = \$110.38$$

If the interest is compounded *monthly,* we have:

$$S = 100 \times (1 + 0.0083333)^{12} = 100 \times 1.10471 = \$110.47$$

And if there is *daily* compounding, the result is:

$$S = 100 \times (1 + 1.000274)^{365} = 100 \times 1.1052 = \$110.52$$

With *simple* interest we would have earned $10.00 for an end-of-year total of $110.00.

These small differences have a way of adding up quickly. You might be surprised at how fast your investment can double. If you have a calculator that does *logarithms,* it is possible to figure this out exactly.

Here's how

DOUBLING YOUR MONEY

The following formula allows you to figure out how long it will take to double your money—any amount of money at any rate of interest—but can only be used with a calculator that has a logarithm button $\boxed{\text{log}}$. This is the formula:

$$n = [\log 2] \div [\log(1+i)]$$

In the formula:

n = the number of interest periods
i = the periodic interest rate divided by 100 (just as in the compound interest formula)
log (1 + i) is obtained by first entering $1\boxed{+}i\boxed{=}$ in your calculator and then pressing the $\boxed{\text{log}}$ button.

Logarithms are never computed by hand and we will not go into their meaning here. They are usually taught in an intermediate algebra or precalculus course. For now—if you intend to do a lot of these computations, buy a scientific calculator that will do both $\boxed{Y^x}$ and logarithms.

EXAMPLE: Let's go back to the formula and try a problem. How many *interest periods* does it take to double your money if the interest rate is 10% compounded quarterly?

SOLUTION: Earlier in this section, we found that i = 0.025 when the interest rate is 10% compounded quarterly. So:

$$1 + i = 1 + 0.025 = 1.025.$$

Next we compute *log* 2 and *log* (1.025)

By calculator:
PRESS 2 $\boxed{\text{log}}$ and
PRESS 1.025 $\boxed{\text{log}}$

You'll find that log 2 = 0.30103 and log 1.025 = 0.01072. Substituting in the formula:

$$n = 0.30103 \div 0.01072 = 28.08$$

It takes 28.08 interest periods (quarters in this case) to double your money. That is just a little more than 7 years. To see this, divide 28.08 by 4 because there are 4 quarters in a year (and 28.08 ÷ 4 = 7.02).

We can check the answer by applying the compound interest formula, $S = P(1 + i)^n$. Suppose your principal is $1,000, the interest rate is 10% compounded quarterly, and the number of interest periods (n) is 7×4 or 28 quarters. Then:

$$S = 1,000 \times 1.025^{28}$$
$$S = 1,000 \times 1.9965$$
$$S = 1,996.50$$

So, in 7 years (28 quarters), we have slightly less than doubled our $1,000 investment. The answer ($1,996.50) is not exactly doubled ($2,000.00) because it takes 7.02 years, not 7 years for money to precisely double.

✓*TIP*   There is a shortcut method to estimating how long it takes to double your money that uses what we'll call *magic numbers*. (These numbers are not really magical, but were derived using advanced mathematical techniques.)

Here's how

MAGIC NUMBERS

We're going to start with an example before we give you the rule of thumb.

When the interest period is quarterly, the way to determine how many years it takes to double your money is to divide the magic number 70 by the quoted interest rate. Suppose the interest rate is 10% compounded quarterly:

$$70 \div 10 = 7 \text{ years to double your money.}$$

That's the same answer we obtained before, using logarithms.

Here's the magic number table:

Table 2
Compound Interest Magic Numbers

IF THE QUOTED INTEREST RATE IS—

AT LEAST:	BUT LESS THAN:	COMPOUNDED:	THE MAGIC NUMBER IS:
3%	5%	Daily, Monthly, or Quarterly	69
3%	5%	Annually	70
5%	20%	Daily	69

IF THE QUOTED INTEREST RATE IS—

AT LEAST:	BUT LESS THAN:	COMPOUNDED:	THE MAGIC NUMBER IS:
5%	20%	Monthly or Quarterly	70
5%	20%	Semi-annually	71
5%	10%	Annually	72
10%	15%	Annually	73
15%	20%	Annually	75

What you do is look up the magic number that corresponds to both the range in which your interest rate falls and to the compounding period. Then you divide the magic number by the exact quoted interest rate and you'll have an estimate of the *number of years* it takes to double your money under those conditions. Using this shortcut magic, the answer is always the number of years.

The answer won't be exact, but in *most* instances it will be accurate to the nearest tenth of a year (± 0.1). In other words, if the "magic answer" is 7 years, the exact answer most probably falls between 6.9 years (-0.1) and 7.1 years ($+0.1$). If you stay within the indicated ranges in the table, you'll never be more than 0.2 (two-tenths) of a year off the exact amount of time it takes to double your money. (Don't forget that 0.2 means "two-tenths of a year" and *not* 2 months. Two-tenths of a year is about 2½ months.)

EXAMPLE: Now, suppose you were offered an interest rate of 9% compounded daily. For this situation, and for *every* situation with daily compounding, irrespective of the rate of interest, the magic number is always 69, so divide 69 by 9. The answer is 7.7 years (that's 7 years and somewhat more than 8 months). If you figure it out with logarithms, you also find the answer to be 7.7 years, rounded to the nearest tenth of a year.

EXAMPLE: You *can* use the magic number table outside of the indicated ranges, but be prepared for less accurate estimates. Let's try such an example. How many years would it take to double your money at 25% compounded annually? We realize that 25% is beyond the values in the table, but using the closest figure, we see that the magic number is 75. Computing, $75 \div 25 = 3$ years. Compare this result with the answer of 3.1 years you obtain by formula ($n = \log 2 \div \log[1 + i]$) and you can see that the two answers are quite close.

✓✓*HOT TIP* Ever hear of the "Rule of 72"? That's a simple investment rule, very much related to our "magic numbers," that lets you determine approximately how long it will take your investment to double in value, *assuming* the earnings (interest) are paid annually and remain in the same account.

To apply the Rule of 72, simply divide 72 by the stated interest rate: if the interest rate is 8%, divide 72 by 8 to find that your money will double in approximately 9 years; at 9%, it will take 8 years (for annual interest rates between 5% and 10%—compare to our compound interest magic numbers table); and so on.

The Rule of 72 and the table of magic numbers are fun and provide quick, shortcut, reasonable estimates to a tantalizing question. But it is a game to play after you've made the more serious decision about where and how to invest your money (to bring the greatest return) or where and how to borrow money (at least cost).

These decisions are *a matter of interest*. And it is *to your interest* to be able to calculate and compare opportunities.

Section 3: . . . and More about Compounding

It's really very hard for people to believe just how much money accumulates as a result of the effects of compound interest—although we saw (Section 2, above) how quickly money can double. Almost everyone can grasp the idea of money earned through simple interest, and, because our examples tend to use small numbers and whole percentages, we can even begin to comprehend how interest compounds over one or two years.

But it is really difficult to conceive how much money you can end up with by just letting your principal and interest continue earning interest for 5 or 10 or 40 years.

EXAMPLE: Your IRA, Keogh, or 401(k) plan are examples of the type of investment where the principal and interest can remain untouched for years and years. (We'll discuss IRA's [Individual Retirement Accounts], Keogh accounts and 401(k) plans in Chapter 3, but for now, remember you don't pay taxes on these earnings until you start making withdrawals.) Suppose you invest $2,000 a year for 10 years. You deposit $2,000 on December 31, 1992, $2,000 on December 31, 1993, $2,000 on December 31, 1994 and so on until December 31, 2001. The question is,

How much would these deposits be worth on December 31, 2002 (one year after the last deposit) if interest is allowed to accumulate and the interest rate is a steady 7% compounded monthly?

Well, let's see.

You've deposited a total of $20,000 over the 10-year period, of which $2,000 compounded monthly for 10 years *and* another $2,000 compounded monthly for 9 years *and* another $2,000 compounded monthly for 8 years and so on . . . *plus* $2,000 that was on deposit for only one year at 7%, compounded monthly. That's . . .

But wait! There's a formula for computing the *total value* of a series of equal investments made over a span of years.

Here's how

COMPUTING TOTAL VALUE

We're going to show you the formula that can be used to compute total value exactly for any size investment, any interest rate, and any number of years. Then, we'll show you a shortcut method for computing how much money you'll end up with for certain specified interest rates and time periods.

The formula for total value (TV) *one year after the last payment is made* is:

$$TV = \frac{P \times A(A^N - 1)}{A - 1}$$

In this formula:

P = The amount of each of your payments
N = The total number of payments
A is computed according to the formula: $A = (1 + i)^k$

In addition:

k = The number of yearly compounding periods
i = The periodic interest rate expressed as a decimal. (That's the given interest rate divided by k and then divided by 100.)

In our example:

P = $2,000 (we made annual payments of $2,000)
N = 10 (10 payments—one each for 10 years)
k = 12 (since interest is compounded monthly and there are 12 months in a year)

$$i = (7 \div 12) \div 100$$
$$= 0.005833$$

Computing A:

$$A = (1+i)^k$$
$$= (1+0.005833)^{12}$$
$$= 1.072286$$

{ To compute any number to the 12th power, use a calculator that has a $\boxed{Y^x}$ or an $\boxed{X^y}$ button: PRESS 1.005833 $\boxed{Y^x}$ 12 $\boxed{=}$

Substituting in the formula we have:

$$TV = \frac{\$2,000 \times 1.072286 \times (1.072286^{10} - 1)}{1.072286 - 1}$$

{ To calculate 1.072286^{10}: PRESS 1.072286 $\boxed{Y^x}$ 10 $\boxed{=}$

$$= \$29,952.24$$

So, the effects of compounding earned us \$9,952.24 in interest (\$29,952.24 less the \$20,000 that we deposited) over a 10-year period.

Now we'll show you a short-cut method.

Here's how

SHORTCUT: TOTAL VALUE

Using the shortcut method to determine how much principal and interest you'll have accumulated after a given number of years of regular investments, you will need a "multiplier." Then you multiply your investment by this number to find total value. Table 3 presents the multipliers that are used to compute total value when interest is compounded *monthly*. Table 4 contains the multipliers to use in cases when interest is compounded or credited *once a year*. And Table 5 shows the multipliers for situations involving *daily* compounding. All three tables cover interest rates from 5% to 20% and terms from 10 to 40 years.

EXAMPLE:

Step 1 First, decide which table is appropriate for your situation.

Step 2 Next, find the multiplier by locating the point at which the percent interest intersects the number of years you have made investments. Let's use the same example: \$2,000 deposited for each of 10 years at 7% compounded monthly.

Table 3
The Accumulation Factor ("Multiplier") for Each $1 of Regular Yearly Investment, with *Monthly* Compounding (12 Compounding Periods Per Year)

NUMBER OF YEARS OF INVESTMENT

PERCENT INTEREST	10	15	20	25	30	35	40
5	13.293	22.882	35.188	50.981	71.249	97.260	130.642
6	14.105	25.030	39.766	59.644	86.455	122.620	171.401
7	14.976	27.426	45.074	70.093	105.561	155.840	227.118
8	15.914	30.101	51.238	82.728	129.644	199.541	303.677
9	16.923	33.092	58.408	98.045	160.104	257.269	409.399
10	18.009	36.438	66.760	116.649	198.731	333.781	555.980
11	19.179	40.186	76.507	139.302	247.871	435.576	760.105
12	20.439	44.387	87.894	166.933	310.523	571.382	1,045.283
13	21.797	49.100	101.218	200.705	390.611	753.114	1,445.082
14	23.261	54.392	116.827	242.049	493.195	996.894	2,007.117
15	24.841	60.339	135.140	292.760	624.896	1,324.770	2,799.537
16	26.545	67.025	156.641	355.031	794.228	1,766.523	3,918.991
17	28.385	74.552	181.925	431.645	1,012.432	2,363.191	5,504.711
18	30.372	83.027	211.674	525.985	1,293.913	3,170.125	7,754.109
19	32.519	92.581	246.730	642.362	1,657.768	4,263.853	10,952.490
20	34.839	103.352	288.060	786.027	2,128.531	5,747.876	15,505.520

Table 4
The Accumulation Factor ("Multiplier") for Each $1 of Regular Yearly Investment, with *Annual* Compounding (1 Compounding Period Per Year)

NUMBER OF YEARS OF INVESTMENT

PERCENT INTEREST	10	15	20	25	30	35	40
5	13.207	22.657	34.719	50.113	69.761	94.836	126.839
6	13.972	24.673	38.993	58.156	83.802	118.121	164.047
7	14.784	26.888	43.865	67.677	101.073	147.914	213.610
8	15.645	29.324	49.423	78.954	122.346	186.102	279.781
9	16.560	32.003	55.765	92.324	148.575	235.125	368.292
10	17.531	34.950	63.003	108.182	180.943	298.127	486.852
11	18.561	38.190	71.265	126.999	220.913	379.164	645.827
12	19.655	41.753	80.699	149.334	270.293	483.463	859.142
13	20.814	45.672	91.470	175.850	331.315	617.749	1,145.484
14	22.045	49.980	103.768	207.333	406.737	790.673	1,529.909
15	23.349	54.717	117.810	244.712	499.957	1,013.346	2,045.954
16	24.733	59.925	133.840	289.088	615.161	1,300.025	2,738.474
17	26.200	65.649	152.139	341.763	757.503	1,668.994	3,667.388
18	27.755	71.939	173.021	404.273	933.320	2,143.652	4,912.600
19	29.404	78.850	196.848	478.431	1,150.389	2,753.919	6,580.510
20	31.150	86.442	224.026	566.377	1,418.258	3,538.009	8,812.628

Table 5
The Accumulation Factor ("Multiplier") for Each $1 of Regular Yearly Investment,
with *Daily* Compounding (365 Compounding Periods Per Year)

NUMBER OF YEARS OF INVESTMENT

PERCENT INTEREST	10	15	20	25	30	35	40
5	13.301	22.901	35.227	51.053	71.373	97.464	130.962
6	14.117	25.062	39.837	59.780	86.700	123.036	172.084
7	14.994	27.476	45.188	70.322	105.987	156.597	228.415
8	15.940	30.178	51.418	83.104	130.376	200.897	306.104
9	16.954	33.186	58.638	98.546	161.121	259.238	413.084
10	18.054	36.578	67.117	117.460	200.453	337.269	562.814
11	19.238	40.381	77.024	140.531	250.596	441.352	771.954
12	20.516	44.650	88.622	168.739	314.713	580.678	1,065.265
13	21.894	49.446	102.218	203.297	396.901	767.728	1,478.006
14	23.384	54.847	118.203	245.777	502.667	1,019.952	2,061.580
15	24.987	60.902	136.916	297.796	638.293	1,358.944	2,884.179
16	26.728	67.760	159.059	362.208	814.231	1,820.019	4,057.983
17	28.610	75.491	185.154	441.674	1,041.720	2,445.328	5,728.608
18	30.646	84.221	215.969	539.953	1,336.668	3,295.886	8,113.835
19	32.848	94.080	252.370	661.571	1,719.407	4,454.048	11,523.440
20	35.224	105.182	295.262	811.720	2,214.970	6,027.685	16,387.070

Looking at Table 3 (interest compounded monthly), we go down the "percent interest" column until we get to 7% and then move one column to the right to "10 years" to find the multiplier. The multiplier is 14.976.

Step 3 Then, to compute total value, multiply your annual investment by the multiplier. In our example, we deposited $2,000 each year, so:

$2,000 × 14.976 = $29,952.00

As you can see, the result of $29,952.00 is the same as the $29,952.24 we got using the total value formula if we round off to the nearest dollar. But note that the shortcut, like the elaborate formula, only works if there was a regular deposit made over a period of years. Also, both methods assume a constant rate of interest.

To show just how quickly compound interest adds up, suppose you begin making regular deposits at age 25 and continue doing so for 40 years. If the interest rate you obtain is 8% compounded daily, the multiplier (from Table 5) is 306.104. This means that you will have over 300 times your annual investment at age 65. For example, depositing $1,000 each year will result in a nest egg of $306,104 at age 65. And you will have invested only $40,000 (40 × $1,000)!

If you had invested $2,000 a year for, say, 35 years, starting when you were 24 years old, and obtained an annual interest rate of 9%, you would have accumulated:

$$\$2,000 \times 235.125 = \$470,250$$

That would be nearly half a million dollars by the time you were 59 years old and almost eligible to withdraw money from your retirement account. Even taking inflation into consideration, that's not a bad return on an investment totalling $70,000!

Hard to believe, but true nevertheless.

3

Investments

Section 1: The Biggest Bang for Your Buck

While April may be the cruelest month, as many taxpayers and readers of T. S. Eliot know, the harshest words in the language of finance may be, "penalty for premature withdrawal." And, indeed, you are harshly penalized for taking your money out of a time deposit account before the account matures.

Most banks call them certificates of deposit (CD's); others refer to them as savings certificates, while still others call them investment certificates. There are seemingly endless variations in names and conditions, but whatever the terminology and terms, these forms of time deposits have certain elements in common. Designed primarily for the smaller investor, these investment options all involve:

1. *A minimum dollar investment,* which can range from a low of $500 to a high of $5,000. While in most banks you almost never can add money to the certificate during its term, you can buy another certificate at any time (provided you satisfy the minimum requirement).
2. *A fixed, predetermined time period (term) during which you have to leave your money on deposit.* The term can vary from 3 months to 5 years: 3-month, 6-month, 12-month, 18-month, and 2-year terms are most usual. Obviously, you have a wide choice.

   In selecting the term that best suits you, you are in effect entering into a *contract* with the bank (or investment house) that guarantees you a fixed rate of interest for the life of the investment. In return, you are committed to leaving your money on deposit for the specified time.

3. *A fixed rate of interest, called the annual interest rate,* which stays in effect for the entire term, irrespective of what happens to interest rates in the economy in general. Note that the interest rate you will get varies with the term of the certificate you choose. In their effort to compete for your investments some banks have taken to "high ball" customers with big, come-on yields that automatically drop after a month or two. For example you might see an ad for a 13-month CD that pays 6.5% for the first 30 days—and then is reduced to 4% for the rest of the term. Being able to figure your annual percentage yield is critical if you want to compare this investment to, say, a simple 12-month CD that pays 5% for the entire term.

4. *Penalties for early withdrawals.* Every time deposit account includes a penalty—in the form of lost interest—for early or premature withdrawal of principal from the account. Except in certain extreme instances (for example, death or adjudicated incompetency of the depositor), the principal may not be withdrawn prior to maturity, in whole or in part, without forfeiting some portion of the interest. Because of these strict and substantial penalties, you should carefully consider whether or not you'll have need of your money before your investment matures when selecting a fixed term investment.

In choosing a CD or savings certificate, your main concerns are the rate of interest and length of the term. If you expect interest rates to go up, you may not want to be locked into a long-term investment; if you expect interest rates to drop, you'll do well for yourself to lock into a high rate for a longer period.

There is a bewildering array of time deposit investment opportunities. One of the primary reasons for choosing this kind of investment is for the security it affords—currently, CD's offer about the same rate of interest that you'd earn in a savings account or a N.O.W. account (a form of interest-bearing checking account); in the 1980's you could have earned quite a bit more interest from a CD. Banks are competing for your business. They offer different interest rates, compounding periods, and terms, all of which make it virtually impossible for the consumer to *easily* figure out the most profitable alternative. But they also help the investor by publishing the "annual effective yield" or annual percentage rate (APR) along with the fixed or annual interest rate.

"What," you ask, "is the annual effective yield? How is it computed?"

Here's how

ANNUAL EFFECTIVE YIELD

The annual effective yield is the actual interest rate you would realize on $100 invested for one year at the quoted compounded interest rate. Since compounding enables you to earn interest on your interest, the actual interest you earn (the annual effective yield) is always higher than the quoted interest rate. The formula for computing annual effective yield takes into account both the quoted interest rate and the compounding period. Thus, the annual effective yield lets you compare the earnings from investments with different quoted compound interest rates. In this respect, it is an indispensable piece of information.

Let's backtrack. When putting money in a savings account, CD, IRA, Keogh or any other savings-investment plan, you are faced with different rates and different compounding periods (as well as different terms). That makes it difficult to compare one option with another.

✓*TIP*   The more frequent the compounding, the greater the increase in earnings.

We already saw that, all other things being equal, compounding increases your earnings. So, if you had to choose between 9.1% simple interest and 9.1% compounded quarterly, you'd select the quarterly compounding. And if the decision was between 9.1% compounded quarterly, and 9.1% compounded daily, you'd know the daily compounding was the better deal.

But in this real world, the options are rarely that simple. More likely, you will be faced with a lower rate that is compounded more frequently than the higher rate: for example, 9.25% compounded quarterly, as contrasted with 9.1% compounded daily. Is the effect of compounding enough to make 9.1% compounded daily a more attractive investment than 9.25% compounded quarterly? That's the kind of question the annual effective yield helps you to answer.

As stated in the beginning of this section, the annual effective yield is the interest you'd earn on $100 if the $100 was left in an account at the quoted interest rate for a period of one year. The formula is:

Annual effective yield $= 100 \ (1+i)^k - 100$

This formula is quite similar to the compound interest formula where:

k = *the number of times per year* that interest is compounded

i = the *periodic interest rate divided by 100* (the periodic interest rate is the quoted interest rate divided by k)

The first part of the formula computes the total interest and principal you would have after one year, based on a principal of $100. By subtracting 100 (the last step in the formula), we remove the principal, leaving the interest earned on $100.

Now, let's compute the annual effective yield on 9.25% compounded quarterly:

k = 4, because interest is compounded quarterly and there are 4 quarters in a year

Periodic interest rate = 9.25% ÷ 4 = 2.3125%, so

i = 2.3125 ÷ 100 = .023125

Substituting in the formula, we have:

$$\text{Annual effective yield} = 100 \ (1 + .023125)^4 - 100$$
$$= 100 \times (1.023125)^4 - 100$$

The quantity $(1.023125)^4$ is read, "1.023125 to the 4th power." That means that 1.023125 is multiplied by itself 4 times.

$$1.023125 \times 1.023125 \times 1.023125 \times 1.023125 =$$

While this multiplication can be done by hand or on your regular calculator, it is best done (faster, without losing your place, and without forgetting the decimal point) on a calculator that computes "powers" directly. Such a calculator, as we described earlier, has a button marked:

$\boxed{Y^x}$ or $\boxed{X^y}$

On *that* calculator, to compute $(1.023125)^4$

PRESS 1.023125 $\boxed{Y^x}$ 4 $\boxed{=}$

The result is 1.0958. This completes the information we need for the formula:

$$\text{Annual effective yield} = 100 \times 1.0958 - 100$$
$$= 109.58 - 100$$
$$= 9.58$$

The result of 9.58 is actually the percentage interest (9.58%) you earn if you invest your money at 9.25% compounded quarterly for a one-year period. The interest earned is exactly the same as it would be if you had deposited your money at 9.58% compounded annually for one year.

Now let's compare the annual effective yield on 9.25% compounded *quarterly* with the annual effective yield on 9.1% compounded *daily*. To do this, we also need to figure out the annual effective yield on 9.1% with daily compounding:

Assuming 365 days per year:

$$k = 365$$
Periodic interest rate $= 9.1\% \div 365 = 0.02493\%$, so
$$i = 0.02493 \div 100 = 0.0002493$$

In addition:

$$\text{Annual effective yield} = 100 \times (1 + .0002493)^{365} - 100$$
$$= 100 \times (1.0002493)^{365} - 100$$

Yes, you *could* multiply 1.0002493 by itself 365 times—but it's almost impossible to do without error. It's also time-consuming and extraordinarily boring. Instead, use a calculator that does "powers":

PRESS 1.0002493 $\boxed{Y^x}$ 365 $\boxed{=}$ 1.0953

To finish substituting:

$$\text{Annual effective yield} = 100 \times 1.0953 - 100$$
$$= 109.53 - 100$$
$$= 9.53$$

Thus, 9.1% compounded daily is the same as 9.53% compounded annually.

This result, 9.53%, is slightly less than the yield of 9.58% obtained on 9.25% compounded quarterly. Therefore, 9.25% compounded quarterly represents a slightly better investment—all other things being equal.

But, as we've seen over and over again, all other things are not equal. There is wide variability in terms and equally far-ranging differences in the

PENALTY FOR EARLY WITHDRAWAL

Banks are not kidding when they talk about the severe penalties you can incur for early withdrawal from a time deposit. Penalties for withdrawing your money from a time deposit account before the date of maturity are always *forfeitures of interest:*

- One typical bank extracts a 3-month interest penalty on time deposits of 6 months or less; it penalizes you 6 months' worth of interest on accounts of more than a 6-month fixed period.
- Another bank requires that you give up 30 days' interest on one year (or shorter) CD's. If has a 90-day interest penalty on time accounts of from 2 to 5 years.

In most (but not necessarily all) cases, the interest penalty is calculated on the basis of simple interest at the quoted rate. We've made up some examples to show you how this is done. To start, let's see what some penalties would amount to on a $500 investment.*

EXAMPLE #1: In the first example, we have a 12-month CD with a quoted interest rate of 9.1% compounded daily. There is a 6-month interest penalty on early withdrawals. Assuming we took out our money after 11 months (330 days), how much is the penalty?

To answer this question, we use the *formula for simple interest* to find out how much interest is earned in 6 months. (Compounding is ignored at this point since we are assuming that penalties are calculated on the basis of simple interest at the quoted rate.)

The formula is:

$$I = Prt$$

Substituting, we have:

$$I = 500 \times .091 \times .5 \quad \begin{cases} 9.1\% = 9.1 \div 100 = 0.091 \\ 6 \text{ months equal six-twelfths of a year,} \\ \text{or } 6 \div 12 = 0.5 \end{cases}$$
$$= 22.75$$

In other words, you have to pay a penalty of $22.75. But your $500 was on deposit for 330 days (about 11 months) at 9.1% compounded daily; applying the compound interest formula, and doing the calculations, you earned $42.87

---

*Because interest rate levels are constantly changing, the example rates used in this book are not meant to reflect current market reality.

in interest. You have to subtract the penalty of $22.75 from the earned interest ($42.87 − $22.75), which means that in 11 months you earned a total of $20.12 on your $500. This would be equivalent to 4.36% compounded daily, which has an annual effective yield of 4.46%—a mighty poor return on your money in comparison to your original expectations. (You'll have to trust us on the calculation of the annual effective yield which calls for more math than you may want to know.)

EXAMPLE #2: Now, let's do another problem. What is a 1-month penalty on $500 invested in a 6-month CD at 9.25% compounded monthly? To obtain the penalty, we need to figure out the simple interest for a 30-day period:

$$I = 500 \times 0.0925 \times 0.083 \qquad \left\{ 1\text{-month} = \frac{1}{12} \text{ year} = 0.083 \text{ year} \right.$$
$$= 3.84$$

The penalty is $3.84. Suppose the $500 was on deposit for 5 months, during which time it earned 9.25% interest compounded monthly. Applying the compound interest formula, $S = P (1 + i)^n$, you can calculate your earnings to be $19.57, from which you must forfeit $3.84. So over the course of 5 months, you actually made $15.73 in interest, which comes out to be the equivalent of 7.46% compounded monthly—or an annual effective yield of 7.72% (again, please trust us on this computation.)

As you can see from these examples, the penalties for early withdrawal of money from time deposits involve substantial percentages. Though the actual dollar amounts in our examples are relatively small, they are not necessarily without meaning. After all, for $22.75 you can get a pretty good restaurant dinner, while for $3.84 you can buy a frozen dinner in the supermarket. Remember, however, that the actual dollar amount of the penalty is proportional to the size of the original investment. If you had invested $5,000 instead of $500, for example, the dollar penalty under the conditions just described would be $10 \times \$3.84 = \$38.40$. That's the price of dinner for two in a modest restaurant!

Typically, people withdraw their principal from investment certificates before the date of maturity because they need the use of the money for other things or because they have a chance to make a better investment. In the first case, if you need your $500 to pay an unanticipated bill, you probably don't have much choice in the matter and you'll just have to accept the loss of interest. If you want to re-invest the principal, however, you will want to know whether the higher rate of return on the new investment will make up for the interest you forfeited. After all, the whole point of investing is getting a bigger bang for your buck.

Section 2: The Bulls and the Bears (Stocks and Bonds)

If you are like most people, you are probably still searching for the one perfect investment that:

- Is Completely *secure* (never worth less than what you paid)
- Is *Liquid* (can be turned back into cash at any time)
- *Appreciates* in value (becomes worth more than its original price)
- Offers a *high yield* (rate of interest)

Unfortunately, investing always involves risks. The more certain you are in one area, such as liquidity, the more you trade-off in another, such as yield.

Stocks rate high in liquidity but vary widely with respect to security and appreciation. For example, utility stocks tend to offer a high dividend yield (which is the ratio of the annual dividend to the price) but generally appreciate very slowly. In contrast, more speculative stocks may show very rapid appreciation—or equally rapid and dramatic drops in value—but are likely to have a zero dividend yield (that is, pay no dividends). Even a well-established corporation can fail to pay expected dividends if its profit status is poor. A company can also declare an unexpected dividend increase when it is doing well.

Typically, the current yield on bonds (which is the ratio of the annual bond coupon payment to the price) is greater than the dividend yield of stocks. However, bonds that have the highest quality ratings (signifying greatest security, according to Standard & Poor's and Moody's), generally have lower yields than bonds that are rated less highly. But no bond is totally secure. In fact, if the corporation, utility, state, or city agency that issues the bond runs into financial trouble and cannot pay its debts (defaults), the bond may not be redeemed at all, or it may be redeemed for less than its face value at maturity. Even in the absence of credit problems, the prices of bonds fluctuate with changing interest rates. Thus, if you sell a bond prior to its maturity, the price you obtain may be higher or lower than the face value of the bond.

The security of some investments—savings accounts and certificates of deposit—is *insured* by the F.D.I.C. (Federal Deposit Insurance Corporation). Some of these investments offer good yields as well, but, except for the addition of interest, they do not appreciate. Those with the highest yields (CD's, for example) may also carry a penalty for early withdrawal.

Liquidity is the most intuitively obvious aspect of investments. It pertains to how readily an investment can be turned back into dollars. You can usually sell a stock or high-quality bond in one business day, although not necessarily for the price you originally paid for it. Not so with real estate (or art, for example, or antiques).

Appreciation tends to be one of the more uncertain factors because it primarily reflects future occurrences and expectations. However, the lure of getting more than your original investment back is one of the major attractions of investing.

The yield on stocks and bonds is a measure of the rate of interest you earn on your original investment.

Here's how

YIELD ON STOCKS

On time deposit investments, for example, savings accounts, the annual effective yield is the rate of interest you'd realize if you left your money on deposit for a 1-year period at the quoted interest rate (see Chapter 3, Section 1).

Stocks are a little different. The dividend yield on stocks is the *amount of money you earn in dividends expressed as a percentage of the price you paid per share of the stock.*

Let's consider the Hilton Corporation stock as an illustration. In December 1991, Hilton Corporation stock was paying a yearly dividend of $1.20 per share. If you owned 100 shares of Hilton, you would earn a dividend of:

$1.20 \times 100 = $120 each year.

And if you owned 500 shares of this stock, your dividends would amount to:

$1.20 \times 500 = $600 a year.

The total amount of dividends you receive is *independent* of what you paid for the stock. It reflects only the *total number of shares you own*. Everyone who owns shares of this stock receives the same per-share dividend. By contrast, the yield on your investment *is* based on the price you paid for each share of stock. It is independent of the number of shares you own. Let's consider your yield to see why this is so. Suppose you bought Hilton on December 25, 1991 at a price of $39.38 per share. The dividend of $1.20 per share (per year) is like earned interest. Finding the yield means finding the interest rate:

Dividend yield = (dividend ÷ price) × 100%

For the shares of Hilton you purchased in December:

Dividend yield = (1.20 ÷ 39.38) × 100%
                = 0.0305 × 100%
                = 3.05%

If your friend had bought the stock for less money, say $38.00 per share, her yield would be higher because the amount of the dividend remains the same:

Dividend yield = (1.20 ÷ 38.00) × 100%
               = 0.0316 × 100%
               = 3.16%

But if another colleague paid more for the stock than you did, his yield would be smaller than yours. If he bought Hilton at $41.25 per share, the yield for him would be:

Percent Yield = (1.20 ÷ 41.25) × 100%
              = 0.0291 × 100%
              = 2.91%

Thus, since the dividend yield on stocks is a percentage based on the dividend and price paid per share, people who buy the same stock at different prices actually earn different yields (just as people who invest in certificates of deposit at different times and through different banks or investment houses can earn different rates of interest).

That's why it's important to distinguish between your yield and the current price of the stock. If a few weeks after you bought Hilton at $39.38 per share, it went up to $45.25 per share, it wouldn't affect your yield (although you would make a significant profit from price appreciation if you sold the stock). For no matter what the current price of the stock is, the fact remains that you invested $39.38 per share, and you earn $1.20 per year on that amount. That's a yield of 3.05% per year during the time you keep the stock, assuming that dividends are paid as promised.

A yield of 3.05% may not seem particularly high. Your money would probably earn a somewhat higher rate of interest in a long-term CD and it would also be guaranteed secure. But you bought Hilton stock anticipating that, in addition to providing a respectable yield, the price would rise. Let's suppose you bought shares at $39.38 and one year later the stock rose to

$46.75 per share. This is an increase of $7.37 per share (a percent increase of 18.72%—see Chapter 1, Section 2.) If you sell the stock at the $46.75 price, you will have realized an *annual return* of 21.77% (21.77% = the 18.72% price appreciation + the 3.05% dividend yield).

If you sell the stock for more than you paid for it, you have a *capital gain* of $7.37 per share. However, if you bought it at $39.38 per share and sold it at $30.35 (a loss of $9.03 a share), you would have a *capital loss* resulting from the price decrease. Irrespective of appreciation (capital gain) or depreciation (capital loss), if the amount of the dividend does not change, the *dividend yield* you realize on your investment also does not change during the time you own the stock. But the annual return varies up or down with capital gains or losses.

✓*TIP*   To compare different investments, for example stocks and CD's, look at the *annual return* and *annual effective yield*. In the previous example, the stock was sold after one year, so computing the annual yield was relatively straightforward. In real life, however, it wouldn't be as simple because you probably won't sell the stock in exactly one year. If you don't, the computation of annual return (called "internal rate of return") becomes very complex, well beyond the scope of this book. You'll need to use a business calculator, such as the Hewlett-Packard 12C, that has the internal programming to compute it automatically.

The current yield on bonds is computed in much the same way as the dividend yield on stocks.

Here's how

CURRENT YIELD ON BONDS

Bonds are issued with a maturity date (the date they are redeemable for the dollar value printed on the bond—the face or par value). Generally, the smallest bond you can buy is for $1,000. Each bond pays a specified rate of simple interest based upon the face value. Your interest payments are obtained in one of two ways. Investors who buy "bearer bonds" clip coupons and send them to the obligor for payment. Typically (but not for all bonds), these coupon payments are made in two semiannual installments. Buyers of "registered issues" automatically receive payments at the appropriate time.

Suppose the coupon rate is 8½% on the bond you own. Applying the simple interest formula (see Chapter 2, Section 2), this means that you receive $4.25 every six months for each $100 of face value:

Interest = one-half of 8½% of $100
= 4.25% of 100

$$= 0.0425 \times 100$$
$$= \$4.25$$

Since the face value of your bond is $1,000, your semiannual interest would amount to $42.50 (10 × $4.25). Over the course of one year, you would receive a total of $85.00 (2 × $42.50) in coupon payments.

EXAMPLE: On November 27, 1991, Georgia-Pacific Corporation issued bonds at 9.5% with a maturity date of 2011. If you bought a $1,000 bond and paid $1,000 for it (that is, paid its full face value), you would earn $95 per year in interest payable in two semiannual installments of $47.50. Your yield would be the quoted 9.5% interest.

But you rarely pay the face value when you buy a bond. Bonds such as those of the Georgia-Pacific Corporation change hands innumerable times before the year 2011 when they can be redeemed at their face value (provided the issuing agency is in sound financial condition). In interim transactions (and even at the time of issue), bonds are usually traded or sold for more or less than their face value. The selling price depends on a variety of factors such as general market conditions, the soundness of the issuing agency, and competing interest rates,

As with stocks *your yield on bonds depends on the price you pay for each $100 of face value*, even though the coupon rate remains unchanged. If you paid a *premium* of more than $100 (per $100 of face value), your yield on Georgia-Pacific Corporation would be less than 9.5%. If you bought the bond at a *discount* (less than $100 per $100 of face value) your yield would be higher than the coupon rate.

Several different calculations are used to describe the yield on a bond investment. One of these yield measures, the *coupon yield*, is the one that is quoted in the newspaper. A second, more complicated (and more accurate) measure is called the *yield-to-maturity*. Let's look at how the coupon yield is calculated.

$$\text{Current Yield} = \frac{\text{Coupon Rate}}{\text{Price paid per \$100 of face value}} \times 100\%$$

On December 26, 1991, Pacific Gas & Electric bonds that mature in 2011 and have a 9⅜% coupon sold at a price of $104.75.
So if you paid that premium price:

$$\text{Current Yield} = \frac{9.375}{104.75} \times 100\%$$
$$= 0.0895 \times 100\%$$
$$= 8.95\%$$

This result (yield) is less than the 9⅜% coupon.

As an example of a discount bond, we found that on December 26, 1991, an AT&T bond, maturing in 2003, sold for $98.375 and had a coupon rate of 7⅛%. In this case:

$$\text{Current Yield} = \frac{7\frac{1}{8}}{98.375} \times 100\%$$
$$= \frac{7.125}{98.375} \times 100\%$$
$$= 0.0724 \times 100\%$$
$$= 7.24\%$$

The current yield of 7.24% was somewhat higher than the coupon rate.

On the same day, another AT&T bond (maturing in 2005) sold for $103.00. It had a coupon rate of 8.8%, but the fact that it was selling at a premium lowered the current yield to 8.54% as you can see:

$$\text{Current Yield} = \frac{8.8}{103.00} \times 100\%$$
$$= 0.0854 \times 100\%$$
$$= 8.54\%$$

The selling price of bonds is, to a great extent, dependent on the prevailing level of interest rates in the marketplace. A bond issued some time ago with a coupon rate that is high by current standards is likely to sell at a considerable premium. The effect of this is to lower the yield to a level that is more in line with current rates.

Like the dividend yield on stocks, the current yield you realize on bonds is independent of what the bond sells for before or after *you* buy it. Your current yield only reflects the price *you* paid and the coupon rate. However, if you sell the bond before maturity, you may experience a capital gain or loss that would affect your total return, just as it did on stocks. Furthermore, if you buy a bond at a discount to face value, when the bond matures you will receive the full face value. Therefore, you will realize a capital gain that represents additional yield on your investment. This additional yield is not considered when you calculate the current yield.

To fully account for the value of all capital gains (and losses if you bought the bond at a premium) when you hold the bond to maturity, investment professionals use a measure called the *yield-to-maturity*. This complicated measure is beyond the scope of this book and requires a special business calculator.

People buy bonds because:

- They tend to be a reasonably secure investment.
- They are redeemable at maturity for their full face value (if the issuer is then financially healthy enough to meet the debt).
- Their yield is relatively high.

Moreover, unlike stocks, some bonds offer additional tax advantages because the earned interest may not be subject to federal or state tax. This is generally true of municipal bonds.

But like any investment, neither stocks nor bonds come up to "par" on every factor that needs to be considered. So, whether you choose to invest in bonds, certificates of deposit, stocks, real estate, or savings accounts, examine your personal needs and weigh the importance of the factors of yield, security, appreciation, and liquidity against one another. Remember, there's no such thing as a perfect investment; investing always involves some gamble. If this wasn't the case, there would be lots more millionaires!

Section 3: On Golden Pond (Tax-Deferred Investments)

Just as a regular schedule of exercise and good eating habits have immediate health benefits as well as payoffs in later years, so too does investing in such tax-deferred basic retirement plans as IRA's, Keogh's, and 401(k) plans. While you must pay taxes when you withdraw the money—the contributions you've made *and* the interest earned on these contributions—the advantages of these plans are that they:

- May save you taxes now
- Offer you a wide choice of investment options
- Earn tax-free interest until withdrawal
- Guarantee you retirement income

It used to be that tax shelters were mainly for the very rich. Now, with government approved IRA (Individual Retirement Accounts) and profit-sharing retirement plans and money purchase pension plans, they are available to many more people, including those with moderate incomes.

Here's how

WHO QUALIFIES?

Everyone who is employed—whether you work for someone else or are self-employed—can open an Individual Retirement Account, irrespective of whether or not you also participate in an employee pension plan. If you have a nonworking spouse, you can also contribute to a spousal IRA. However, recent changes in the Tax Reform law affect the *deductibility* of your IRA contribution.

The deductibility of your IRA contribution depends on several factors, including:

- Your marital status and whether or not you and your spouse actually live together;
- Whether you are an "active participant" in an employer-sponsored retirement plan—that is, a company retirement plan;
- Whether you file as single, married filing jointly, or married filing separately on your federal tax return; and
- The amount of your adjusted gross income.

Taxpayers should check the rules carefully to see if they qualify for a deduction.

Irrespective of whether your IRA contribution is deductible, each year, you may contribute up to $2,000 or 100% of your *earned* income, whichever is less. If you are the only wage earner with a nonworking spouse, you may shelter an additional $250 in separate IRA's (for an annual total of $2,250). However, if both you and your spouse are employed, each of you can open an IRA of your own, so that a working couple's annual contribution can be as much as $4,000. And it doesn't matter whether you file a joint return!

The amount of money you contribute to an IRA, of course, is up to you. Even if you or your spouse is covered by a company plan, you can take the full IRA deduction as long as your adjusted gross income is less than $40,000 on a joint return or $25,000 for singles. If it's deductible in full or in part, partial deductions obtain if you're covered by a company plan and your adjusted gross income is between $40,000 and $50,000 (joint) or $25,000 and $35,000 (single). It is in your best interest to try to invest the maximum each year. From a tax perspective, it's better to invest something than nothing and many banks and investment institutions will allow you to open some type of IRA with an initial deposit of as little as $50. They also permit subsequent contributions of as little as $10.

You can start a new IRA every year (or contribute to an already established one). You also have a wide choice of investment vehicles and institutions. In fact, in any one year you may open as many different IRA's as you like as long as your total investment for the year does not exceed the

$2,000 maximum. And you have until April 15 to add to an already established IRA or open an IRA for the previous calendar year—and to make your contributions to it. (Note that if you file on a fiscal year basis, your deadline may be different. Check it with an accountant.)

Everyone can have an Individual Retirement Account—which may or may not be deductible. Note, however, that even if your contribution is partially or completely nondeductible, the earnings on this money are not taxed until you withdraw the funds. New tax laws also allow sole proprietors, partners, or employees to invest in other types of retirement programs, such as a Profit Sharing Retirement Plan (PSRP) or Money Purchase Pension Plan (MPPP)—both Keogh-type plans. They can also invest in SEP or Simplified Employee Pension plans and can contribute to these types of accounts as long as they have self-employment income, regardless of their age. You need not run a full-time business to set up a Keogh plan; you qualify if you have income from consulting, free-lancing, or a sideline business. And you can have a Keogh plan in addition to an IRA. We advise you to talk to an accountant and to your employer to determine whether you're eligible to participate in these. If you qualify for these Qualified Retirement Plans, you can contribute up to 25% of "earned income" as defined by law for a MPPP, and 15% for a PSRP and SEP Plan up to a maximum of $30,000. (Because of changes in the definition of "earned income," the 25% and 15% contribution always works out to 20% or approximately 13.043%, respectively.) If you can, contributing to an IRA and to another tax-deferred retirement plan is doubly advantageous.

The deadline for opening a *first* tax-deferred basic retirement plan is December 31. Once opened (even with a small investment—say, $100), however, you can keep investing each year's contribution (up to the maximum) until April 15 of the following year (or the deadline, including extensions, for filing your federal tax return).

If you qualify, these contributions are deductible, meaning you defer paying income taxes on the money you contribute. This is a savings in real dollars now.

Here's how

TAX DEDUCTIONS

To find out how these tax-deferred basic retirement plans save you taxes now, let's assume that your contribution meets the deductibility criteria. This means that the amount you invest in an IRA, SEP, Keogh, or 401(k) plan each year is deducted from your taxable (net earned) income for the year, even if you don't itemize your deductions. This results in your paying less income tax for the years you make investments.

Let's suppose you are a married taxpayer filing a joint return and your taxable income this year is $55,000. Using the 1991 Federal Tax Rate Schedule (reprinted in Chapter 1, Section 3, Table 3, and discussed there), we see that this income puts you in the 28% tax bracket. This means that according to Schedule Y-1 you would owe in federal taxes:

$$\$5,100 + 28\% \text{ of } (\$55,000 - \$34,000)$$
$$= \$5,100 + (0.28 \times \$21,000)$$
$$= \$5,100 + \$5,880$$
$$= \$10,980$$

% means hundredths, so
$28\% = 28 \div 100 = 0.28$.
Of means "times."

By calculator:
PRESS .28 ☒ 21,000 ▱ ⊞ 5,100 ▱

A $2,000 tax-deferred contribution to a retirement plan reduces your taxable income from $55,000 to $53,000. Recomputing your taxes on this basis, you will obtain:

$$\$5,100 + 28\% \text{ of } (\$53,000 - \$34,000)$$
$$= \$5,100 + (0.28 \times \$19,000)$$
$$= \$5,100 + \$5,320$$
$$= \$10,420$$

We find that your tax bill is now $10,420—a savings of $560 in taxes $10,980 − $10,420) because of your tax-deferred contribution. (Since you are in the 28% tax bracket, you could have computed your tax saving directly as 28% of $2,000. That's $0.28 \times \$2,000 = \$560$.)

Had you invested *less* than $2,000, you would still realize tax savings amounting to 28% of your contribution as long as your total income fell into this same tax bracket. What happened if your tax-deferred contribution places you in a *lower* tax bracket? Let's answer that question with another example.

Suppose your taxable income was $35,500 before you made the $2,000 tax-deferred investment. That means you were in the 28% tax bracket, according to Schedule Y-1. On this income, the total taxes are:

$$\text{Total Taxes} = \$5,100 + 28\% \text{ of } (\$35,500 - \$34,000)$$
$$= \$5,100 + 0.28 \times \$1,500$$
$$= \$5,100 + \$420$$
$$= \$5,520$$

Your $2,000 investment lowers your taxable income to $33,500 which moves you into the 15% bracket. On this income, the total taxes are:

Total Taxes = 15% of $33,500
    = 0.15 × $33,500
    = $5,025

So you saved $495 in taxes

✓*TIP*   If you want to estimate your savings on your $2,000 contribution in one step, look at where the last $2,000 fell in the tax table. Most of it was in the 28% range and $500 was in the 15% range. A quick calculation of the tax savings is 28% of $1,500 ($420) plus 15% of $500 ($75), or a total of $495. This is precisely the actual savings.

Exactly *what* do you contribute *to* when you contribute to an Individual Retirement Account or 401(k) plan, or one of the other tax-deferred retirement plans? You have several options, and your contributions can be invested in many ways.

Here's how

INVESTMENT OPTIONS

To begin with, you are not limited to investing through a bank or savings institution. You can also invest through a mutual fund, insurance company, or brokerage firm. Not all types of institutions necessarily offer the full range of investment opportunities, but generally you can also select the type of investment you prefer. In the case of 401(k) plans, companies generally offer a menu of investment options including stocks, bonds, and money market accounts. First decide what investment alternative—mutual funds, CDs, etc.—you want to pursue. Then pick an appropriate financial institution based on rates, convenience, past experience, reputation, and so on. We found the books, *How to Buy Stocks,* by Engel and Boyd (7th Ed.), Bantam Books, 1983, and *The New York Times Complete Guide to Personal Investing,* by Gary L. Knott, Times Books, 1987, to be particularly helpful introductory guides.

In deciding which type of investment you want and which institution you want to manage your money, you need to consider yields, growth, and appreciation as well as fees and commissions.

Banks and savings institutions typically offer money market or time deposit IRA's. Banks and savings institutions generally don't charge a fee for opening a retirement plan. However, they may impose some annual maintenance charges (which can range from $5 to $20 per year), and some banks have fees for transferring your account, for closing it out, or for distributing the funds to you when you retire. Almost all banks and savings institutions charge a fee (of about $25) for early withdrawals from a time deposit investment like a CD, in addition to the interest penalty they levy.

In general, mutual fund institutions, insurance companies, and brokerage firms charge various fees and/or commissions for every transaction. There may be a start-up fee, an annual fee, an administrative fee, a management fee, a commission on each year's contribution, and/or a commission on securities bought and sold. Since fees and commissions can mount up substantially and thus cut into the interest you earn, it's a good idea to consider the fee structure when you are comparing investment options—as well as the factors of security, liquidity, and appreciation.

Now, let's examine one of the most common investments: time deposits. We've already described some of the important features of time deposits—lengths of term, yields, and penalties for early withdrawals—when we covered certificates of deposit in Section 1 of this Chapter. Now, we will discuss variable and fixed interest rates.

Banks offer both short- and long-term growth accounts, typically of 6 months to 5 years duration. Each has either a fixed or variable rate of interest.

A *fixed rate* account guarantees a specific rate of interest for the entire term. The fixed rate in one representative bank reflects the most current 52-week yield on U.S. Treasury Notes. If you think interest rates have peaked and will be going down, or if you just feel more comfortable knowing *exactly* by how much your money will grow, the fixed rate option is right for you. So, lock in the current market rate for 6 months, 18 months, 30 months, or 5 years!

Variable interest rates change, usually monthly. The interest you earn keeps pace with fluctuating market conditions and may be based on the latest 3-month average of the yields of 1-year U.S. Treasury Bills. Variable rate time deposit accounts are for investors who believe interest rates are going up. (They can, of course, go down.)

You can split your contribution, placing a portion in a fixed rate account and the remainder in a variable rate one, or splitting it between time deposits and stock market mutual funds. This lets you enjoy some of the gains of an appreciating stock market but gives you some cushion against extreme losses because not all of your money is in stocks. But remember, you can always transfer your tax-deferred investment to a new kind of investment (or to a new trustee). In the case of an IRA, Keogh, or SEP, you can transfer to a new trustee within 60 days of closing an account without losing your tax-deferred advantage. Nor do you lose your tax-free interest.

(Remember, if you move a CD before it matures, you still have to pay an early withdrawal penalty.)

Here's how

TAX-FREE INVESTMENT

The interest on your investment in a tax-deferred retirement plan is tax-free as long as you leave it in the plan or transfer it speedily (within 60 days) into a new plan. Thus sheltered, your account grows rapidly as the interest compounds over time.

To show just how quickly it can grow, let's look at a simple example of a tax-deferred account that earns a constant 8% interest rate compounded annually.

In the beginning of the first year, you make a tax-deferred contribution of $2,000. By the end of the year, you earned $160 interest (8% of $2,000—see Chapter 2), so that you have a total of $2,160.

By the end of the second year, you earn 8% interest on the first year's contribution (that's 8% of $2,160 or $172.80). You also earn another $160 on the second year's contribution of $2,000 (which you made at the beginning of the second year).

By the end of the second year, you have earned:

$172.80 interest on $2,160 (the first year's investment plus one year interest), and $160 interest on $2,000 (the second year's investment).

Adding principal and interest together, your funds now total $4,492.80.

At the beginning of the third year (which is the same as the end of the second year), you deposit an additional $2,000. For the third year you earn 8% interest on $6,492.80 ($4,492.80 + $2,000). That's $519.42 in interest, so at the end of the third year, you have $6,492.80 + $519.42 or $7,012.22.

If you continued contributing $2,000 for each of 10 years (that's $20,000) at a fixed 8% compounded annually, the principal and accumulated interest would amount to $31,290.97—not one penny of which has been taxed! (A shortcut way of doing this computation will be explained in the next chapter.)

If you deposited $2,000 a year in a nonsheltered account, your net earnings would have been much less (unless, of course, the interest rate was astronomical), because you would be taxed on the interest earned. Also, by contributing to a tax-deferred retirement plan, if you qualify, you also save in taxes by reducing your taxable income each year for 10 years. So there's really a dual advantage to this method of saving.

Can you really get away with *never* paying taxes on this money? The answer is "no," but the retirement plans were designed so that you generally pay a lower tax rate on the money when you do pay taxes.

Here's how

RETIREMENT INCOME

Let's suppose that you withdraw the $31,290.97 in one lump sum as soon as you are able to without penalty (at age 59½—more about this later). If you were still working and reporting a taxable income of $20,000 at that time, the addition of $31,290.97 gives you a taxable income of $51,290.97 for that year, immediately moving you up into the 28% tax bracket. Your federal income taxes for the year (again using the 1991 Schedule Y-1) would be $10,142.97.

$$= \$5,100.00 + 28\% \text{ of } (\$51,290.97 - \$34,000.00)$$
$$= \$5,100.00 + 0.28 \times \$17,290.97$$
$$= \$5,100.00 + \$4,841.47$$
$$= \$9,941.47$$

However, let's assume that at age 59½ you withdraw the whole $31,290.97, but you have no other taxable income. You are, in effect, retired. The entire $31,290.97 falls in the 15% tax bracket, and your retirement income is taxed at a lower rate.

But you can still earn income *and* reduce the taxes that you will have to pay on your retirement income because:

- You don't have to withdraw your money in one lump sum.
- Nor do you have to do it beginning at age 59½.

You *may* begin to receive distributions from your tax-deferred account as early as age 59½ (even if you are still working). You *must* begin to make at least the required minimum withdrawal by April 15 of the year after you reach 70½. You also may not make regular contributions to a tax-deferred retirement account during or after the tax year in which you reach age 70½, but if you have a non-working spouse under age 70½, you can contribute up to $2,000 each year to a spousal IRA. (Isn't there something slightly incongruous about half-birthdays after your fifth or sixth one?)

You can opt for a lump sum payout or withdraw your money *in install-ment*. Taxes are paid *only* on the amount of money you withdraw, and you continue to earn tax-free interest on the balance remaining in the account. Withdrawals from a tax-deferred account are taxed immediately as if they had been earned during one tax year.

The fact is that IRA's and the other tax-deferred plans we've referred to are intended to be *retirement plans*. The assumption is that when you make withdrawals you will pay less tax on the money because your annual income will be less than it was during the years when you were making your contribution to the plan. When your income is less, you are in a lower tax bracket and your taxes are less.

You may choose to withdraw all or part of your retirement account funds at any time before age 59½. But, if you do so for reasons other than because you are totally and permanently disabled, you may have to pay a *penalty tax*, usually a percentage of the amount withdrawn—in addition to income taxes, of course. And, if you withdraw from an IRA time deposit investment that has not yet matured, you will also incur an early withdrawal penalty and may have to pay an early withdrawal fee.

There are many good reasons for opening an IRA account and other tax-deferred retirement plans (if you are eligible) and for trying to make the maximum annual contribution, if possible. Among the more compelling reasons are that such plans can en*rich* your later years and make them truly *gold*en.

RETIREMENT ANNUITIES

After years of careful planning and saving for your future, you see that future—retirement—fast approaching. Because you want to be assured a reasonable standard of living for the rest of your life, you will not withdraw all of your savings at once. Instead, you will take out just enough each year to meet your needs. This will allow the balance of your savings to continue to grow in accordance with the rate of return your investments provide.

But how to figure out how much to take out of savings?

Here's how

Level Payment Annuities

Table 1 can be used to help you estimate how much you will need to finance your retirement. For example, suppose that you estimate that, in addition to Social Security and your company retirement benefits, you will need an additional $10,000 per year to maintain a comfortable standard of living. We hate to be morbid, but it is also necessary to guess how many years of payments you want to be assured of. If you retire at 65, you probably do not expect to live for another 50 years, but today's life expectancies are such that it is fairly likely that you will be around for another 10 or 20 years.

If at retirement, you invest $98,181 at 8% annually, you will be able to withdraw $10,000 per year for each of the next 20 years, beginning one year from today. At the end of that time, your account balance will be totally depleted. The value $98,181 can be found in Table 1 by reading across the row corresponding to an 8% interest rate and down the column corresponding to 20 years of payments.

At first it may seem surprising that you can withdraw a total of $200,000 ($10,000 per year for 20 years) by investing only a little more than $98,000. To understand how this initial investment can be so productive, keep in mind that the money that is not withdrawn continues to earn 8% interest, thereby replenishing some of the money that you do withdraw each year.

As another example of the use of Table 1, continue reading across the row corresponding to an 8% return. Observe that it takes an initial investment of $112,578 to finance 30 payments of $10,000. Also, if you are able to invest $125,000 at 8%, then you (or your heirs) can withdraw $10,000 per year forever.* It is interesting to observe that the incremental investment

Table 1
Initial Investment Required to Realize Annual Payments of $10,000

INTEREST RATE	5	10	15	20	25	30	PERPETUAL
3.00%	45,797	85,302	119,379	148,775	174,131	196,004	333,333
4.00%	44,518	81,109	111,184	135,903	156,221	172,920	250,000
5.00%	43,295	77,217	103,797	124,622	140,939	153,725	200,000
6.00%	42,124	73,601	97,122	114,699	127,834	137,648	166,667
7.00%	41,002	70,236	91,079	105,940	116,536	124,090	142,857
8.00%	39,927	67,101	85,595	98,181	106,748	112,578	125,000
9.00%	38,897	64,177	80,607	91,285	98,226	102,737	111,111
10.00%	37,908	61,446	76,061	85,136	90,770	94,269	100,000
11.00%	36,959	58,892	71,909	79,633	84,217	86,938	90,909
12.00%	36,048	56,502	68,109	74,694	78,431	80,552	83,333
13.00%	35,172	54,262	64,624	70,248	73,300	74,957	76,923
14.00%	34,331	52,161	61,422	66,231	68,729	70,027	71,429
15.00%	33,522	50,188	58,474	62,593	64,641	65,660	66,667
16.00%	32,743	48,332	55,755	59,288	60,971	61,772	62,500

YEARS OF PAYMENTS

---

*Our examples do not take into account the effects of taxes. Typically, investments in a retirement account grow on a tax-deferred basis. This means that you only pay taxes when you

that is required to go from 30 years of payments to perpetual payments is not much greater than the incremental investment that is required to go from 20 years of payments to 30 years.

Finally, going back to Table 1, we note that the cost of funding your retirement income decreases dramatically as the rate of return increases. If your money is invested at a 3% interest rate, it takes $148,775 to fund 20 years of $10,000 payments. But notice that the cost of funding your retirement drops by more than 50%, to only $62,593, when the rate is 15%.

Table 1 can be used to determine the cost of financing any level of annual payments you desire. For example, if you want to withdraw $30,000 per year, simply multiply the results in the table by 3, because $30,000 is 3 times the $10,000 base level. As an illustration, with an 8% interest rate, you would need an initial investment of $294,543 (3 × 98,181) to provide you with an annual income of $30,000 for 20 years.

Inflation-Protected Annuities

The examples we have just given you do not reflect the effects of inflation on retirement income. If prices do not change, $30,000 may be sufficient to meet next year's needs. However, if the costs of food, rent, transportation, and clothing increase substantially over time, the buying power of $30,000 will increase proportionately. Over the past 65 years, inflation has averaged about 3.2% annually. In the early 1980s, inflation was a very high 10% to 12%. In 1993, the inflation level was comparable to the lower historical averages.

When planning for retirement, it is necessary to protect yourself against the ravages of inflation by assuming that your annual payouts increase by the same rate as prices.

Here's how:

Suppose we estimate that long-term inflation will be about 3% per year. If our first payout is $10,000, the second should be $10,300 ($10,000 plus 3% of $10,000), the third should be $10,609 ($10,300 plus 3% of $10,300), and so on. Table 2 shows the cost of funding just such a payment stream for the same range of interest rates and payout periods as in Table 1. For example, reading across the row corresponding to an 8% interest rate and down the column corresponding to 20 years of payments, we see that it takes $122,500, to fund a payout of $10,000 one year from now, followed by a series of 19 additional annual payments that increase at 3% per year.

---

withdraw money. Thus, you may earn 8% on your investment without paying taxes, but you will have to pay taxes at the appropriate rate on your withdrawals of $10,000 or whatever amount you withdraw each year.

70 The Only Math Book You'll Ever Need

Table 2
Initial Investment Required to Realize Annual Payments Starting at $10,000 and Growing with Inflation (Inflation Rate = 3%)

ANNUAL INTEREST RATE	YEARS OF PAYMENTS						
	5	10	15	20	25	30	PERPETUAL
3.00%	48,544	97,087	145,631	194,175	242,718	291,262	ERR
4.00%	47,161	92,098	134,916	175,714	214,589	251,630	1,000,000
5.00%	45,839	87,476	125,296	159,648	190,851	219,193	500,000
6.00%	44,574	83,188	116,638	145,615	170,718	192,463	333,333
7.00%	43,363	79,205	108,830	133,317	153,556	170,284	250,000
8.00%	42,204	75,501	101,773	122,500	138,854	151,757	200,000
9.00%	41,092	72,053	95,380	112,956	126,198	136,176	166,667
10.00%	40,026	68,837	89,576	104,505	115,250	122,985	142,857
11.00%	39,003	65,837	84,297	96,998	105,735	111,746	125,000
12.00%	38,022	63,033	79,485	90,307	97,426	102,109	111,111
13.00%	37,079	60,410	75,090	84,326	90,138	93,795	100,000
14.00%	36,174	57,953	71,067	78,962	83,716	86,578	90,909
15.00%	35,303	55,650	67,378	74,137	78,033	80,278	83,333

For comparison, we recall that it took "only" $98,151 to fund 20 annual payments of $10,000 each. The additional $24,349 in required investment ($122,500 − $98,151) can be viewed as the "cost of inflation."

Reading across to the last column of Table 2, we see that it takes $200,000 invested at 8% annually to finance annual payments (starting at $10,000) that grow at 3% forever. Note that if your investment only earns the 3% inflation rate, it is impossible to finance perpetual payments—no matter how much you invest, eventually all the funds will be depleted.

Table 2 can also be used to find the cost of funding different levels of payouts by simply multiplying the amounts in the table just as we did in Table 1. For example, 10 years of payments that grow at 3%, starting with a $25,000 payout requires an initial investment of $188,752.50 if the interest rate is 8%. This result was calculated by multiplying the table amount ($75,501) by 2.5 because $25,000 is 2.5 times $10,000.

To further illustrate the costs of inflation, we are including one more table that shows the cost of financing 20 years of inflation-sensitive payments at different inflation rates and different interest rates. Here's how to use Table 3. Reading across the first row, we see that with a 6% interest rate, the cost of financing 20 annual payments starting with an initial payout of $10,000, assuming a consistent annual 3% inflation rate, is $145,615. This cost rises by almost $13,000 to $158,399 if the inflation rate is 4%, and by another $14,000 if the rate of inflation is 5%. At higher interest rates, the initial financing costs are lower but they still increase significantly at higher inflation rates.

Table 3
Initial Investment Required to Realize 20 Annual Payments Starting at $10,000 and Growing with Inflation

INTEREST	INFLATION RATE		
RATE %	3.0%	4.0%	5.0%
6.00	$145,615	$158,399	$172,689
8.00	122,500	132,475	143,580
10.00	104,505	112,384	121,121
12.00	90,307	96,607	103,563

Living comfortably in retirement is very costly, especially in an economy in which inflation is a fact of life. When armed with a better understanding of tomorrow's needs, we can come to a reasonable balance between our need to save for tomorrow and our desire to spend and enjoy today.

4

Long-term
Loans

Section I: Home Mortgages, Automobile Loans and
Present Value

A home mortgage is probably the largest loan you'll ever have, and the next largest is a car loan; these are generally a family's biggest lifetime purchases. But any type of loan where you pay back part of the principal and interest in *periodic installments* over time (or have this paid out to you, as with a pension) works much the same way.

Until the introduction of adjustable-rate mortgages (home mortgages where the interest is adjusted periodically, typically reflecting yields on U.S. Treasury bills or some other index of interest rate levels), all mortgages used a fixed rate of interest. This meant that the interest rate did not fluctuate and, consequently, the amount of the payment remained the same over the life of the loan.

The amount of the payment is based on the total amount of the loan, the length of the loan (its *term*), the rate of interest, and the frequency of payments. It also involves the concept of *present value,* which we will be discussing later. Did you ever wonder how the bank figures it out given all the different combinations?

Here's how

COMPUTING LOAN PAYMENTS

There is a remarkable formula for computing the exact amount of your mortgage payment or the payment on an auto loan or on any other long-term, fixed-rate loan. (There are also tables that list the exact payment for different size loans at different rates of interest.) The formula is remarkable for two reasons: first, because it can be used for computing payments on loans of any amount, at any fixed interest rate, and for any term; second, because, although it looks imposing,

$$\text{Payment} = \text{Amount of loan} \times \frac{i \times (1+i)^n}{(1+i)^n - 1}$$

it has only two unknowns, n and i, and is very similar to the compound interest formula with which we worked in Chapter 2.

In this formula:

i = *The periodic interest rate divided by 100.* (The periodic interest rate is the quoted interest rate divided by the number of times per year that interest is compounded.)
n = *The total number of payments to be made.*

Let's use the formula to figure out the mortgage payments on a house you're thinking of buying. Suppose the house costs $125,000, and you want to make a $15,000 down payment. This means that you must borrow $110,000, so you shop around and find a 30-year mortgage at 9% interest with monthly payments. (When loans are paid off monthly, the quoted interest rate is usually understood to be compounded monthly.) The question that you are interested in is, "How much will each monthly payment be?"

The first step in applying the formula is to figure out the unknowns:

n, the total number of payments to be made, is 30 (years) × 12 (months per year) = 360;
The *periodic interest rate* is 9% ÷ 12 (the number of times per year that interest is compounded) = 0.75%,
So i, the periodic interest rate divided by 100, is $0.75 \div 100 = 0.0075$
And the quantity $(1+i)^n$ is

$$(1+i)^n = (1 + 0.075)^{360}$$
$$= (1.075)^{360}$$

To figure this out, you must have a calculator that computes powers. If you don't have one, you'll have to buy one because it's the only way to do the arithmetic needed for finding exact mortgage payments (and for figuring out compound interest in Chapter 2). Look for a calculator that has a $\boxed{Y^x}$ or $\boxed{X^y}$ key. Then, to compute $(1.0075)^{360}$ on this calculator,

PRESS 1.0075 $\boxed{Y^x}$ 360 $\boxed{=}$

(The answer will take a moment to appear; it takes the calculator a few seconds to complete this computation.)

$$(1+i)^n = (1.0075)^{360} = 14.73058$$

Now we can substitute in the formula:

$$\begin{aligned}
\text{Payment} &= \$110,000 \times \frac{0.0075 \times 14.73058}{14.73058 - 1} \\
&= \$110,000 \times \frac{0.11048}{13.73058} \\
&= \$885.09
\end{aligned}$$

This means that your mortgage payment would be $885.09 a month, every month for 30 years.

Now let's see what happens to the monthly payment on a $110,000 *20-year* mortgage at 9%. (We've changed the *term* of the mortgage, but left all of the other variables the same.)

Going back to the formula:

n now equals 20 (years) $\times$ 12 (months per year) = 240;
$i = (9\% \div 12) \div 100 = 0.0075$;
$(1+i)^n = (1+0.0075)^{240} = (1.0075)^{240}$

PRESS 1.0075 $\boxed{Y^x}$ 240 $\boxed{=}$ 6.00915

Substituting in the formula, we have:

$$\begin{aligned}
\text{Payment} &= \$110,000 \times \frac{0.0075 \times 6.00915}{6.00915 - 1} \\
&= \$110,000 \times \frac{0.04507}{5.00915} \\
&= \$989.73
\end{aligned}$$

Thus, the effect of a shorter term (all other factors being unchanged) is to *increase* the amount of each payment.

EXAMPLE: To figure out the payments on a car loan we use exactly the same formula. You are going to buy a new car that costs $15,000 by paying $3,000 and borrowing the remainder ($12,000). The quoted rate is 10%, and you are to make equal monthly payments for 3 years. How much will your monthly car payment be?

Looking at the payment formula, we find the amount of the loan to be $12,000. The number of payments, n, is 36; i, the periodic interest rate is $(10\% \div 12) \div 100 = 0.00833$. And $(1 + 0.00833)^{36} = (1.00833)^{36}$

PRESS 1.00833 $\boxed{Y^x}$ 36 $\boxed{=}$ 1.34802

So:

$$(1 + i)^n = 1.34802$$

Substituting in the formula, we have:

$$\text{Payment} = \$12,000 \times \frac{0.00833 \times 1.34802}{1.34802 - 1}$$
$$= \$12,000 \times \frac{0.01123}{0.34802}$$
$$= \$387.18$$

> Keep as many decimal places as possible. Keeping only a limited number of decimal positions results in slight inaccuracies in your answers compared with the answers the bank would obtain. In this case, for example, the bank would have computed the payment to be $387.21.

This is the exact amount of your monthly installment.

You can always compute the exact payment on any fixed-rate loan by using the formula. But it is also possible to *estimate* the payment and to be within ± 3% of the exact amount.

Here's how

ESTIMATING LOAN PAYMENTS

✓*TIP* We are going to show you a new technique for approximating payments on long-term loans. The technique assumes *monthly payments* (and monthly compounding of interest). You can do it on a regular calculator.

In estimating payments you will be working with a formula that has two numbers for you to compute. The *approximation formula* is:

Approximate payment = (Amount of loan) × (*i*) × (Multiplier)

In this formula, *i* is an old friend:

i = the periodic interest rate divided by 100. (In this case, the periodic interest rate is the quoted interest rate divided by 12 because interest is compounded monthly.)

The *multiplier* is found according to the value of N in the following table:

Table 1
Estimated Loan Payment Multipliers

N (YEARS OF LOAN × QUOTED INTEREST RATE)		MULTIPLIER
Over 400	⟶	1.00
300–400	⟶	1.03
250–299	⟶	1.08
200–249	⟶	1.12
175–199	⟶	1.17
161–174	⟶	1.21
150–160	⟶	1.26

As an illustration, a 25-year mortgage at 12% has a multiplier of 1.03 because 25 × 12 (years of loan × interest rate) = 300 and the multiplier for the 300–400 range is given as 1.03 in Table 1 above.

Let's redo the example of a house purchase that we did earlier in this chapter. That example involved a 30-year, $110,000 mortgage at an interest rate of 9%. Now, we want to figure out the *approximate* monthly payment.

$i = (9\% \div 12) \div 100 = 0.0075$

Again, when computing i, keep as many decimal places as appear on your calculator display. Rounding off can lead to serious over- or under-approximations when you multiply i by a very large number (the value of the mortgage).

N = 30 (years of loan) × 9 (the quoted interest rate) = 270
From Table 1, 270→Multiplier of 1.08.

Substituting, we have:

Approximate payment = $110,000 × 0.0075 × 1.08 = $891.00

(The exact payment was found to be $885.09.)

Now, let's do a few problems assuming monthly compounding of interest.

PROBLEM 1:

What is the approximate monthly payment on a $90,000 mortgage at 10.5% for 15 years?

$i = (10.5 \div 12) \div 100 = 0.00875$
$N = 15 \times 10.5 = 157.50 \rightarrow$ Multiplier (from Table 1) of 1.26
Appropriate payment = $90,000 × 0.00875 × 1.26 = $992.25

(The exact payment is $994.86.)

PROBLEM 2:

You borrowed $120,000 for 30 years at an interest rate (compounded monthly) of 8.5%. About how much will this loan cost you a month?

$i = (8.5 \div 12) \div 100 = 0.00708333$
$N = 30 \times 8.5 = 255 \rightarrow$ Multiplier = 1.08
Approximate payment = $120,000 × 0.00708333 × 1.08 = $917.99

(The exact payment is $922.70.)

PROBLEM 3:

Our friends, the Wests, have a $60,000 mortgage that they got some time ago at 7% interest. If they had 30 years to pay off the loan, about how much is their monthly payment?

$i = (7 \div 12) \div 100 = 0.005833333$
$N = 30 \times 7 = 210 \rightarrow$ Multiplier = 1.12
Approximate payment = $60,000 × 0.005833333 × 1.12 = $392

(The exact payment is $399.)

In talking about mortgages and other long-term loans, we are talking about payments spread over a period of time. However, a payment made today is *worth more* to the bank than next month's payment because the bank can re-invest today's payment and begin to earn interest on it immediately. This is the concept of *present value*. The exact payment formula (and also the approximate one) for mortgages takes the concept of *present value* ("today's worth") into account.

Here's how

PRESENT VALUE: TODAY'S WORTH

Let's spend a little time exploring how banks, financial institutions, and many, many people think about the *value of money*. Consider what happens first to a small and then to a large amount of money in a relatively short period of time. Start with $1,000 and a going rate of interest of 8% compounded monthly. (If you haven't done so, please read the section about compound interest in Chapter 2; it will help you better understand the concept of present value.)

Eight percent compounded monthly is 0.67% per month $(8\% \div 12 = 0.67\% = 0.0067)$. Therefore, the interest earned on $1,000 in one month is:

$$0.0067 \times \$1,000 = \$6.67$$

So, if you had invested your $1,000 for a month, you would have earned $6.67. If instead of investing it you kept the money under the proverbial mattress or in a no-interest checking account for one month, you'd "lose" $6.67 in the sense that you *could have* earned that $6.67 had you invested your money at that interest rate.

Would not having the $6.67 disturb you very much? Maybe yes or maybe no, depending on various factors. In fact, it might be worth $6.67 to you to have the convenience and joy of sleeping with $1,000 under your mattress for a month!

But $1,000 is a small sum of money in comparison to the $5,000,000 you just won in the lottery. Here, one month's interest earns:

$$0.0067 \times \$5,000,000 = \$33,500$$

This is more than many people make in a year. You certainly won't want to lose a month's interest on $5,000,000 for the fun of taking it to bed with you!

To emphasize the impact of interest on large sums of money, let's change the interest rate to 7% compounded *daily*. This means:

The daily interest rate $= 7\% \div 365$

$\phantom{\text{The daily interest rate}} = 0.019178\%$ $\left\{ \begin{array}{l} \% \text{ means hundredths, so} \\ 0.019178\% = 0.019178 \div 100 \\ = 0.00019178 \end{array} \right.$

$\phantom{\text{The daily interest rate}} = 0.00019178$

O.K., what does this daily interest rate yield on $1,000?

$0.00019178 \times \$1,000 = \$0.19$

That's about 19¢ per day. But on $5,000,000?

$0.00019178 \times \$5,000,000 = \$958.90$

That's about $959 a day!

And on one billion dollars (that's $1,000,000,000), 7% compounded daily, yields:

$0.00019178 \times \$1,000,000,000 = \$191,780$

That's $191,780 each day!

Banks have millions or possibly even billions to invest each day. With such large sums involved it is understandable that a bank would never want to leave its money idle (that is, uninvested)—not even for a single day. So, when a bank gives you a loan for a house, or car, or anything else, it must consider the *time* that each payment is to be made. A payment made today is worth more to the bank than a payment made next month because the bank can reinvest today's payment immediately.

Thus, a payment made to the bank far into the future is worth much less than the payment made today. *The worth today of a payment made in the future is called present value.*

Let's start with an example of 10% simple interest. If you start out investing $100 today at 8% simple interest, you will have $108 in your account at the end of one year. In other words, given a 8% simple interest rate, $108 a year from today has a *value* of $100 today. To put it still another way, the *present value* of $108 a year from today is $100 (at the quoted interest rate).

EXAMPLE: Now, let's do an example involving compound interest. Suppose you deposit $500 in an account paying 9% compounded monthly. That means you earn 0.75% (0.0075) interest each month. At the end of the first month you will have 0.75% interest credited to your account:

$0.0075 \times \$500 = \$3.75$ interest credited to your account,
for a total of $503.75

(The present value of $503.75 one month from today at 9% compounded monthly is $500.)

Since for the second month and each month following, you'd be earning interest on the principal and on the previous month's interest, you'd want to employ the compound interest formula to compute, for example, how much money you'd have at the end of six months. Repeating the compound interest formula, $S = P (1 + i)^n$, where:

P = the principal or amount with which you start
(In the example, P = $500.)
i = the periodic interest rate divided by 100
(The periodic interest is the quoted interest rate divided by the number of times per year that interest is compounded. In our example, it is $9\% \div 12 = 0.75\%$, and $i = 0.75 \div 100 = 0.0075$.)
n = the number of interest periods
(There are 6 interest periods in our example because interest is credited monthly for 6 months.)
S = the final amount of money, including both principal and interest.

To finish the problem, substituting in the formula:

$$S = \$500 (1 + 0.0075)^6$$
$$= \$500 (1.0075)^6$$
$$= \$500 \times 1.0459$$
$$= \$522.95$$

So, at the end of 6 months, you'd have $522.95 in your account.

Another way of saying this is, given 9% interest compounded monthly, $522.95, 6 months from today has a present value of $500 (is worth $500 today).

Up to now, we have actually been taking an amount of money and determining its *future value* and then restating the example in present value terms. But we can also answer the question, "At the quoted interest rate, how much money do I need to deposit today to pay a $400 bill due 3 months from now?" Answering this type of question involves *discounting*, which is finding the present value of a future amount.

Here's how

DISCOUNTING

The essence of a problem in discounting is that the final amount is known (that's S, the interest plus principal in the compound interest formula), and you wish to find the original amount (P, or principal in the compound interest formula). By way of illustration, suppose you know that the first year of your oldest child's college tuition will be $7,500. That bill must be paid two years from now. If the interest rate is now 8% compounded monthly, how much money will you need to put into a bank account today to ensure that you have $7,500 in two years? In this example, the $7,500 is like a payment made in the future, and the amount you'd need to deposit today (equivalent to the Principal in the formula) is like the present value of that payment.

Now, the compound interest formula can be restated as:

$$FV \text{ (future value)} = PV \text{ (present value of payment)} \times (1+i)^n$$

If the future value or payment is a known quantity, say our $7,500 tuition, the question being asked by this formula is, "What amount deposited today (PV) will yield, with accumulated interest, the given future value (FV) of $7,500 after n (24 in this case because it must be paid in two years or 24 months) interest periods?" The equation above can be solved for present value (PV) by dividing both sides of the equation by the quantity $(1+i)^n$. This results in the formula for present value:

$$PV = \frac{FV}{(1+i)^n}$$

In our example, we have:

$$PV = \frac{\$7,500}{(1+0.00667)^{24}} \left\{ \begin{array}{l} 8\% \text{ compounded monthly gives a periodic} \\ \text{interest of } (8 \div 12) \div 100 = 0.00667 \end{array} \right.$$
$$= \$7,500 \div (1.00667)^{24}$$
$$= \$7,500 \div 1.1729$$
$$= \$6,389.50.$$

The present value of $7,500 two years (24 months) from today, given an interest rate of 8% compounded monthly, is $6,389.50.

EXAMPLE: Let's try another example. Suppose that in settlement of an old debt you are promised $1,000 three years from now. If the going rate of interest is 6% compounded quarterly, how much should you be willing to accept today as an equivalent payment of the debt?

In the problem just posed, the future payment is \$1,000; i is $(6\% \div 4) \div 100 = 0.015$; and $n = 12$ (3 years × 4 quarters per year). Using your calculator, you will find that:

$$(1 + i)^n = (1 + 0.015)^{12}$$
$$= 1.1956$$

Therefore:

$$PV = \$1,000 \div 1.1956 = \$836.40$$

Therefore, \$836.40 today is worth \$1,000 three years from now if the interest rate is 6% compounded quarterly.

We can double check the answer by applying the usual compound interest formula:

$$S = P \, (1 + i)^n$$

We just found that, if the interest rate is 6% compounded quarterly, and $n = 12$, then $(1 + i)^n = 1.1956$. With $P = \$836.40$, we have:

$$S = \$836.40 \times 1.1956 = \$1,000.00$$

Knowing how discounting works lets you do advance planning to meet future obligations. We just saw that an obligation of \$1,000 three years from now is equivalent to an out-of-pocket expense of \$836.40 today (assuming the interest is 6% compounded quarterly). With larger obligations and longer time spans, the difference between present and future value is even more dramatic. For example, a tuition bill of \$20,000 due in 15 years only requires a deposit of \$4,788 today if the interest rate is 10% annually. If the interest rate was only 7% annually, that same future tuition bill would require a substantially higher deposit (of \$7,249) today. Long-range planning and an understanding of present value can ease the burden of future obligations.

There is a special type of security, called a zero coupon bond, that can be particularly helpful in funding future obligations. The prices of such bonds are based on the present value concept.

Here's how

ZERO COUPON BONDS

In the previous chapter (Chapter 3, Section 2), we talked about bonds that make periodic interest payments based on the contractual "coupon rate." For example, a 10-year 8% bond with a $1,000 face value will pay the bearer of the bond $80 (8% of $1,000) per year in interest. Typically, but not universally, that interest, or "coupon payment," is paid in semiannual installments of $40. At the end of 10 years, the bondholder receives the final coupon payment and the $1,000 face value.

Investors holding such "coupon" bonds, are continually faced with the decision of what to do with the coupon payments, or interest. If you bought such a bond with the intention of using the coupon payments as a regular source of income, then you will simply spend the money. On the other hand, you might have bought the bond as a long-term investment to finance, say, the first year of your child's college education. In this case, you will want to reinvest the coupon payments as soon as they are received. Because interest rates change over time, the rate you earn on your reinvested coupons will almost assuredly be quite different than the yield you expected when you purchased the bond.

For investors who do not need the coupon income as immediate income and who wish to avoid reinvestment problems, *zero coupon bonds* offer an attractive alternative. Such bonds do not pay coupons. Instead, they are sold at a "discount" from the face value that is paid at maturity. The difference between the maturity value and the purchase price can be viewed as the interest that is paid over the life of the bond. As long as you hold the bond to maturity, the interest rate or "yield-to-maturity" is assured.

Zero coupon bonds may be issued by corporations, government agencies, or municipalities, or they may be direct obligations of the U.S. government. In the last case, the securities are called Treasury STRIPS and there is no risk of default because of government backing.

It's important to remember, as we have said, that there are no perfect investments and Treasury STRIPS are no exception. As long as you hold the bond, you will not receive coupon income, but you will have to pay federal taxes on the implicit interest that is earned each year. In addition, if for some reason you should have to sell the bond prior to maturity, you will be subject to price fluctuations due to interest rate changes. For perfectly legitimate, but highly technical reasons, the price of a zero coupon bond is much more sensitive to changes in interest rates than the price of a coupon bond with the same maturity.

As an example of how zero coupon bonds are priced, suppose that you want to be buy a bond with a $10,000 face value, due in 18 years. Assuming that the current interest rate (or yield) for 18-year zeros is 8% compounded semiannually, we can calculate the discounted purchase price of the bond with the present value formula.

$$PV = \frac{\$10,000}{(1+0.04)^{36}} \quad \left\{ \begin{array}{l} \text{8\% compounded semiannually gives a} \\ \text{periodic interest of } (8/2) \div 100 = 0.04 \end{array} \right.$$

$$= \$10,000 \div (1.04)^{36}$$
$$= \$10,000 \div 4.1039$$
$$= \$2,436.71$$

In this example, you can buy an 18-year zero coupon bond with a face value of $10,000 for $2,436 and change.

For the 1992–93 academic year, tuition, room and board at a good college was about $23,000. It is shocking, but not unreasonable to assume that those costs will more than double over the next 18 years. In the 2010–11 academic year, college costs may total $50,000. And that's for only *one* year of school! To fund that first year of college for your newborn you might buy zero coupon bonds with a face value of $50,000, five times the face value of the bond in the example above. Assuming the same interest rate, the cost of those bonds will be

$12,183.55 or 5 × $2,436.71.

If interest rates are higher when you buy the bond, the cost of the same face value of bonds will be lower. Paying a lower price for the same future value actually means that you get more interest. By the same token, when interest rates are lower than 8%, the cost of the bonds will be greater.

Section 2: It's Real Interest-ing: Determining Your Interest

In negotiating long-term loans and especially the type of "easy payment" installment plans that are advertised in some tabloids, you may at times be receiving information that actually disguises the true interest rate you will be paying.

Here is an example of what we mean. When buying a car, you may be told that your loan for $8,000 has been approved and that you'll be paying it off in 36 equal monthly payments of $270, with interest amounting to $1,720.

$36 \times \$270 = \$9,720$ (in total payments)
$\underline{\quad -\$8,000}$ (the amount of the original loan)
$= \$1,720$ (in interest)

On the surface, it looks as if you'll be paying an interest rate of about 7.2%, figured like this:

($1,720 ÷ $8,000) ÷ 3 years = 7.17% or about 7.2% per year

But *that's the wrong way to compute the rate of interest!* Surprisingly, when you do it correctly, you'll find you're actually paying 13.93%.
Here's how

ESTIMATING INTEREST RATES

In our discussion of home and automobile mortgages (Chapter 4, Section 1), we explained the concept of present value—today's worth of a payment made in the future. You will recall from that discussion that as far as the bank or lending institution is concerned, since each payment is made at a different time, it is worth a different amount. Given the number of payments, the amount of each payment, and the amount of the loan, it is possible to compute the interest rate on a long-term loan. But it is not possible to come up with an exact formula for that interest rate. However, advanced mathematics has made it possible to develop computer programs that will compute interest rates to any desired degree of accuracy.

And that's what we did in this Section.

✓✓*HOT TIP*   By writing a computer program, we developed a table of payment factors that lets you look up your *annual effective interest rate* (the actual yearly interest rate you are paying, as we showed in Chapter 3, Section 1) to the nearest whole percent. For all practical purposes, this close approximation is all you need to be sure you're not being charged a usurious rate of interest. (By definition, a usurious rate is a rate greater than that allowed by law.)

The technique that we devised shows you a brand new way of determining your annual effective interest rate, given the amount of the loan, the size of the equal monthly payments, and the total number of payments. This technique, published in this book, by us, for the first time, involves first applying a very easy formula, then looking the answer up in the table that ends this section. That's all there is to it.

Let's start.

In order to estimate the rate of interest you are paying, you must know the amount of the loan, the monthly payment (assuming the loan is paid off in equal monthly installments), and the total number of payments. Then work through three steps, as follows:

Step 1 Compute the *Payment Factor* (PF):
$$PF = \frac{\text{Amount of loan}}{\text{Monthly payment}}$$

In the example above:

$$PF = \frac{8,000}{270}$$

$$= 29.63$$

Step 2 Knowing the total number of payments (36 in the example), locate this number in the left column of Table 2. Parts A, B, and C of this table are for interest rates between 7% and 11%, 12% and 16%, and 17% and 21%, respectively. The first column lists the number of monthly payments, in 12-months intervals, ranging from 12 (one year) to 360 (30 years). (Parts D and E of Table 2 apply to rates of interest greater than 21% and list monthly payments in 3-month intervals). If the exact number of payments you are required to make is not given to you by the advertiser or seller, use the nearest number, but remember that your resulting estimate will be somewhat less precise.

Step 3 After you have located the total number of payments in the appropriate column, look across that row horizontally from the left to right to find the column where the PF closest to the one you have computed appears.

Notice that the PFs decrease as you go across the row and that the last number in this row in Part A of the table is 30.37. So, go on to Part B of Table 2 and continue moving across the row that corresponds to 36 payments.

You will find the number 29.98 followed by 29.60, which is as close to our example PF of 29.63 as we can get using these tables. Now, look up to the top of the column in which 29.60 appears, and you'll see the heading corresponds to an interest rate of 14%. Since 29.60 isn't exactly 29.63, the interest rate you'll be paying is not exactly 14% but close to it. That's almost twice the original estimate of 7.2% we thought we'd be paying when we did the obvious, but incorrect, calculations at the beginning of this section.

To check this approach to estimating interest rates, let's go back to one of the examples we used in the previous section, even though we know the interest rate. There, we were buying a $125,000 house with a $15,000 down payment. We borrowed $110,000 on a 30-year mortgage at a 9% compounded monthly (this is an annual effective rate of 9.38%). In that section, we learned to compute the monthly payment, which amounted to $885 (rounded off from the exact amount of $885.09).

The *incorrect way* of determining interest rates would be:

$885 × 360 (that's 12 payments/year × 30 years)
= $318,600 (total interest and principal to be paid)
− $110,000 (in principal)
  $208,600 (in interest)

Dividing 208,000 by 110,000 and then by 30 years results in an (incorrect) interest rate of 6.32%. Note again that this interest rate is *not correct* (even though the arithmetic we did is accurate) because it does not take the timing of the payments into account.

To obtain a more *correct* and reasonable estimate, we first need to find the PF:

$$\frac{\$110,000}{\$885} = 124.29$$

Going down the left column of Table 2 to 360, we look across the row to the PF closest to 124.29. The numbers in that row in the table are 153.63, 139.68, 128.29, 118.22, etc. Thus, the closest number to 124.29 is 128.29. Looking up this column that this number appears in to the heading at the top, we see 9%—fairly close to the 9.38% (effective rate) we know we are paying. In fact, if you wanted to be a bit more precise, you could observe that the actual PF, 124.29, is not quite halfway between the table values corresponding to the 9% and 10% rates. Therefore, you would estimate the effective rate to be a little less than 9.5%.

EXAMPLE: As another example, suppose that an installment loan on a complete 7-piece bedroom set costing $3,400, calls for $30 down and 24 equal payments of $179. Fortunately, such "bargains" are much less common than they used to be. This "bargain" ends up costing you $4,326, of which $926 is interest. In this case, we find the PF to be:

$$\frac{\$3,400 - \$30}{\$179} = 18.83$$

A value close to this value is found in Part E of Table 2 along the row corresponding to 24 payments. The value of 18.89 in the table is in the column representing a 27% annual effective rate of interest!

Truth in Lending Laws make it easier to find out the rate of interest you are actually paying if you bother to read the whole contract. The table in this section provides you with a fast way of double-checking annual effective interest rates under the condition of equal monthly payments . . . It's in your interest to do this check.

Table 2
Table of Payment Factors (Part A)

TOTAL NUMBER OF PAYMENTS	7%	8%	9%	10%	11%
		ANNUAL EFFECTIVE INTEREST RATES			
12	11.57	11.51	11.46	11.40	11.29
24	22.38	22.17	21.97	21.76	21.37
36	32.49	32.04	31.61	31.19	30.37
48	41.93	41.18	40.46	39.75	38.41
60	50.76	49.64	48.57	47.54	45.59
72	59.01	57.48	56.02	54.62	51.99
84	66.72	64.74	62.85	61.05	57.72
96	73.93	71.45	69.12	66.90	62.82
108	80.66	77.67	74.86	72.22	67.38
120	86.95	83.43	80.14	77.06	71.46
132	92.84	88.76	84.98	81.45	75.09
144	98.33	93.70	89.42	85.45	78.34
156	103.47	98.27	93.49	89.08	81.24
168	108.27	102.51	97.22	92.38	83.82
180	112.76	106.43	100.66	95.38	86.13
192	116.95	110.06	103.80	98.11	88.20
204	120.87	113.42	106.69	100.59	90.04
216	124.53	116.53	109.33	102.85	91.68
228	127.96	119.41	111.76	104.90	93.15
240	131.16	122.08	113.99	106.76	94.46
252	134.15	124.55	116.04	108.46	95.63
264	136.94	126.83	117.91	110.00	96.68
276	139.55	128.95	119.63	111.40	97.61
288	141.99	130.91	121.21	112.67	98.44
300	144.28	132.73	122.66	113.83	99.19
312	146.41	134.41	123.99	114.88	99.85
324	148.40	135.97	125.21	115.84	100.45
336	150.26	137.41	126.32	116.71	100.97
348	152.00	138.74	127.35	117.50	101.45
360	153.63	139.98	128.29	118.22	101.87

Table of Payment Factors (Part B)

TOTAL NUMBER OF PAYMENTS	12%	13%	14%	15%	16%
		ANNUAL EFFECTIVE INTEREST RATES			
12	11.29	11.24	11.19	11.13	11.03
24	21.37	21.18	21.00	20.82	20.46
36	30.37	29.98	29.60	29.23	28.52
48	38.41	37.77	37.16	36.56	35.41
60	45.59	44.67	43.78	42.92	41.30
72	51.99	50.77	49.59	48.46	46.33

Table of Payment Factors (Part B, *continued*)

TOTAL NUMBER OF PAYMENTS	12%	13%	14%	15%	16%
		ANNUAL EFFECTIVE INTEREST RATES			
84	57.72	56.16	54.68	53.27	50.63
96	62.82	60.94	59.15	57.46	54.31
108	67.38	65.17	63.08	61.10	57.45
120	71.46	68.91	66.51	64.26	60.14
132	75.09	72.22	69.53	67.01	62.43
144	78.34	75.15	72.18	69.41	64.39
156	81.24	77.74	74.50	71.49	66.07
168	83.82	80.04	76.54	73.30	67.50
180	86.13	82.07	78.32	74.87	68.73
192	88.20	83.86	79.89	76.24	69.77
204	90.04	85.45	81.27	77.43	70.67
216	91.68	86.86	82.47	78.46	71.43
228	93.15	88.11	83.53	79.36	72.09
240	94.46	89.21	84.46	80.15	72.65
252	95.63	90.18	85.27	80.83	73.12
264	96.68	91.05	85.98	81.42	73.53
276	97.61	91.81	86.61	81.93	73.88
288	98.44	92.49	87.16	82.38	74.18
300	99.19	93.08	87.64	82.77	74.43
312	99.85	93.61	88.07	83.11	74.65
324	100.45	94.08	88.44	83.40	74.84
336	100.97	94.50	88.76	83.66	75.00
348	101.45	94.86	89.05	83.88	75.13
360	101.87	95.19	89.30	84.07	75.25

Table of Payment Factors (Part C)

TOTAL NUMBER OF PAYMENTS	17%	18%	19%	20%	21%
		ANNUAL EFFECTIVE INTEREST RATES			
12	11.03	10.98	10.93	10.89	10.79
24	20.46	20.29	20.12	19.96	19.64
36	28.52	28.18	27.84	27.52	26.89
48	35.41	34.86	34.33	33.82	32.83
60	41.30	40.53	39.79	39.07	37.70
72	46.33	45.33	44.37	43.44	41.70
84	50.63	49.40	48.22	47.09	44.97
96	54.31	52.85	51.45	50.13	47.65
108	57.45	55.77	54.17	52.66	49.85
120	60.14	58.25	56.46	54.77	51.65
132	62.43	60.34	58.38	56.53	53.13
144	64.39	62.12	59.99	57.99	54.34
156	66.07	63.63	61.35	59.21	55.34

Table of Payment Factors (Part D)

TOTAL NUMBER OF PAYMENTS	ANNUAL EFFECTIVE INTEREST RATES				
	22%	23%	24%	25%	26%
168	67.50	64.91	62.49	60.23	56.15
180	68.73	65.99	63.45	61.08	56.82
192	69.77	66.91	64.25	61.79	57.36
204	70.67	67.68	64.93	62.37	57.81
216	71.43	68.34	65.49	62.87	58.18
228	72.09	68.90	65.97	63.27	58.48
240	72.65	69.37	66.37	63.62	58.73
252	73.12	69.77	66.71	63.90	58.93
264	73.53	70.11	66.99	64.14	59.09
276	73.88	70.40	67.23	64.33	59.23
288	74.18	70.65	67.43	64.50	59.34
300	74.43	70.85	67.60	64.63	59.43
312	74.65	71.03	67.74	64.75	59.51
324	74.84	71.18	67.86	64.84	59.57
336	75.00	71.30	67.96	64.92	59.62
348	75.13	71.41	68.04	64.99	59.66
360	75.25	71.50	68.11	65.04	59.69

Table of Payment Factors (Part D)

TOTAL NUMBER OF PAYMENTS	ANNUAL EFFECTIVE INTEREST RATES				
	22%	23%	24%	25%	26%
3	2.90	2.90	2.89	2.89	2.89
6	5.66	5.65	5.64	5.62	5.61
9	8.29	8.26	8.24	8.21	8.18
12	10.79	10.75	10.70	10.66	10.61
15	13.17	13.10	13.04	12.97	12.90
18	15.44	15.34	15.25	15.16	15.07
21	17.59	17.47	17.34	17.22	17.11
24	19.64	19.48	19.33	19.18	19.03
27	21.59	21.40	21.21	21.03	20.85
30	23.44	23.22	23.00	22.78	22.56
33	25.21	24.95	24.69	24.43	24.18
36	26.89	26.59	26.29	26.00	25.72

Table of Payment Factors (Part E)

TOTAL NUMBER OF PAYMENTS	ANNUAL EFFECTIVE INTEREST RATES				
	27%	28%	29%	30%	31%
3	2.88	2.88	2.88	2.87	2.86
6	5.60	5.59	5.57	5.56	5.54
9	8.16	8.13	8.11	8.08	8.03
12	10.57	10.52	10.48	10.44	10.36
15	12.84	12.77	12.71	12.65	12.53
18	14.98	14.89	14.80	14.72	14.55
21	16.99	16.88	16.77	16.66	16.44
24	18.89	18.75	18.61	18.47	18.20
27	20.68	20.50	20.34	20.17	19.85
30	22.36	22.16	21.96	21.76	21.38
33	23.95	23.71	23.48	23.25	22.81
36	25.44	25.17	24.91	24.65	24.15

Section 3: It's in the Cards: Credit Card Interest

When you buy by credit card, using VISA or MASTERCARD or a department store charge, you are, in effect, borrowing money. Just as in any other situation that involves loans, you, as the borrower, must pay interest.

Not all credit card interest is the same. Different companies (banks, stores) charge different rates, and they compute the balance on which the interest is paid in different ways, with considerable variation from one to another. So, shopping for a credit card makes as much sense as trying to get the best price on any deal. Since all cards require you to make payments each month when there is a balance due, getting the "best" card—the one with terms that are most advantageous to you—will result in a saving every month.

The credit card company's terms are usually stated in small print in an out-of-the-way place on the bottom or on the back of your monthly statement. Here's a typical example:

We [the credit card company] figure the finance charges [a fancy phrase for interest] on your account as follows: On purchases, by adding together the outstanding purchase balance at the end of each day in the billing cycle, including unpaid finance charges and giving effect to a purchase from the later of the date the purchase is made or the first day of the billing cycle, dividing that sum by the number of days in the billing cycle and applying to that balance the periodic rate.

This 77-word sentence is confusing, to say the least. Notice that the rate of interest that you are being charged is referred to in the last three words—"the periodic rate." The rest of this long sentence describes how the company computes the balance on which the interest is applied. In trying to understand what you will be paying in finance charges, you need to consider two factors: first, the *annual effective interest rate* and second, the *average daily balance*.

What is *not* important is to spend any time checking the bank's monthly calculations since these are done by computer. It's safe to assume that all the arithmetic done by computer is correct. You should, however, check to make certain that there were no errors in recording the individual charges and the cash advances—you do this by matching the amount shown on your statement with the amount on your (the "customer's") copy of each charge slip.

The rest of this section is about how finance charges work. Let's start with the rate of interest you actually pay.

Here's how

ANNUAL EFFECTIVE INTEREST RATE

The annual effective rate of interest is based on the quoted rate of interest (also given on each monthly statement you receive from the credit card company) and is the actual interest rate you pay per year for the use of the credit card.

To understand the annual effective rate, think of the bank, store, or credit card company as earning interest—from you. Interest can be charged both on new purchases and on old balances. If you waited until the end of the year to pay your bill, you would be adding *additional interest* on the interest that you pay on purchases and on the balances. In effect, you would be paying interest on the interest.

This is exactly what happens with many, but not all, credit cards. The charges include any previous balance owed, plus new purchases, plus interest. Some time elapses between the time the charges are accrued and the time you get your statement. Consequently, you may be billed for interest on the interest you owe from the time of the previous period until the time your payment is received.

Interest paid on interest, the situation we just described, is exactly the same as the concept of compound interest (which we talked about in Chapter 2, Section 2). This concept also underlies the idea of *annual effective yield*, or annual effective *rate*. If you're uncertain about the relationship between annual effective yield (or annual effective rate) and the quoted interest rate, rereading Chapter 3, Section 1 will help.

The formula for annual effective rate is:

Annual effective rate = $100 (1+i)^k - 100$

In this formula:

> k = the number of *times per year* that interest is paid;
> i = *the periodic interest rate divided by 100*. (The periodic interest rate is the quoted interest rate divided by k.)

Notice that now we need to know more about the periodic interest rate and the quoted interest rate.

The Sonesta Bank's interest charges on its MASTERCARD purchases are given as follows (this information appears on the monthly statements):

> Annual percentage rate 18.90%
> Monthly periodic rate 1.575%

The *monthly periodic rate* is determined by dividing the annual percentage rate (or quoted interest rate) of 18.90 by 12 months, as follows:

$$\frac{18.90}{12}$$
$$= 1.575\% \text{ per month}$$

This is the rate of interest you pay every month on your average daily balance. (This rate is "the periodic rate" that was referred to in the long sentence we quoted at the beginning of this section.)

Going back to the formula for annual effective rate, let's first finish computing i:

$$i = \frac{1.575}{100} = 0.01575$$

And $k = 12$.

Therefore, substituting in the formula we have:

$$\text{Annual effective rate} = 100 \times (1 + 0.01575)^{12} - 100$$
$$= 100 \times (1.01575)^{12} - 100$$

Computation of 1.01575^{12} must be done on a calculator that has a $\boxed{Y^x}$ or an $\boxed{X^y}$ button. To do this computation:

PRESS 1.01575 $\boxed{Y^x}$ 12 $\boxed{=}$

The result is 1.2063, so:

$$\begin{aligned}
\text{Annual effective rate} &= 100 \times 1.2063 - 100 \\
&= 120.63 - 100 \\
&= 20.63
\end{aligned}$$

This means that the quoted annual percentage rate of 18.90% results in your paying interest at the rate of 20.63% per year!

In some sense, determining the annual effective rate was the easy part. Now, let's consider how the Sonesta Bank figures your average daily balance.

Here's how

AVERAGE DAILY BALANCE

We'll show you two different methods for computing the average daily balance and we will compare them. Before we start, remember that the *average daily balance* is not calculated in the same way by all banks or credit card companies. Many variations are possible. For example, some companies exclude finance charges if you pay your balance on the required date, while others include them. Some do not add interest charges on to new purchases *if* those purchases are paid for on time. There are credit cards that require you to pay interest from the date the purchase is made (the date of the transaction), while other cards begin computing interest from the *posting date* (the day the company or bank received the bill from the establishment where the purchase was made).

As you can begin to see, the possibilities and combinations are so varied that we couldn't cover them all here. Rather, we present two typical examples of different methods of computing the average daily balance on a real bill we just received from the Sonesta Bank.

Method 1

This is a reproduction of our December/January monthly statement:

POSTING DATE	TRANSACTION DATE AND DESCRIPTION		$ AMOUNT
12/14	12/07	New England Store	28.88
12/27	12/17	Dynasty Szechuan	29.70
01/07	12/21	Sunshine Records	19.20
01/07	12/27	Razzi Restaurant	29.16
01/09		Payment—Thank you	200.00

Here's some additional information we collected from this bill and from the previous month's statement:

Closing Date:	1/11 (from current bill)
Previous Closing Date:	12/11 (from previous bill)
Previous Balance (including interest due):	$1,291.47 (from current bill)
Prior Month's Interest Charge:	$20.48 (from previous bill)
Monthly Periodic Rate:	1.575% (from both bills)

Method 1 of calculating the average daily balance is the simplest method and also the one with the terms that are most advantageous to you. It *excludes* both interest payments and new purchases, but takes into account the fact that your payment (of $200) wasn't received until 01/09. It also takes into consideration that $1,291.47 (the previous balance including $20.48 in interest) was due on 12/11. Method 1 is most likely to be used by credit card companies that want you to pay your full balance each month.

Three steps are involved in computing the average daily balance:

Step 1 Compute the number of days in the billing cycle that runs from 12/11 to 01/11

From 12/11 to 12/31 there are:	20 days
From 12/31 to 01/11 there are:	11 days
TOTAL	31 days

Step 2 Subtract the interest due from both the previous balance and from the payment (Method 1 excludes interest payments):

net previous balance due = $1,291.47 − $20.48 = $1,270.99

net payment made = $200.00 − $20.48 = $179.52

Step 3 Find the average daily balance:

(a) Count the net previous balance due (from Step 2) 31 times (there are 31 days in the billing cycle):

$1,270.99 \times 31 = 39,400.69$

(b) Count the net payment made (from Step 2) 3 times (since the payment was "present" during only 3 days of the billing cycle—i.e., the payment was made on 01/09 and the closing date is 01/11 = 3 days)

$179.52 \times 3 = 538.56$

(c) Subtract the amount calculated in (b) from amount calculated in (a) and divide by 31
$$(39,400.69 - 538.56) \div 31 = 1,253.62$$

This ($1,253.62) is the average daily balance using Method 1.

To compute *current interest due* charges, multiply the average daily
balance by the specified monthly periodic rate:

Current interest due $= 1.575\% \times \$1,253.62$
$= 0.01575 \times \$1,253.62$
$= \$19.74$

To compute the *total amount of money you owe* on your Sonesta Bank
credit card this month, add the previous balance due (including previous
interest due), the current interest due, the total of all new purchases (which
are $28.88 + $29.70 + $19.20 + $29.16 = $106.94), and subtract the amount
paid:

Total due $= \$1,291.47 + \$19.74 + \$106.94 - 200.00$
$= \$1,218.15$

According to this method, our total amount due is $1,218.15.

We will now figure out what we would owe using another common
method of computing the average daily balance. This method, Method 2, is
more costly to you.

Method 2

This method for computing the average daily balance tends to be fa-
vored by stores and banks that encourage you to carry forward a balance by
requiring only a minimum payment from you each month. As you might
expect, Method 2 includes both interest charges and new purchases in the
computation of the average daily balance.

Method 2 was somewhat cryptically described in the 77-word sentence
that started this section. According to this sentence, new purchases will be
counted from the *transaction date* unless that date is earlier than the beginning of the current billing cycle. If that is the case, the new purchase will
be considered to have occurred on the first day of the billing cycle. (This
situation can arise if the establishment where the purchase was made did not
submit its bills to the credit card company quickly.)

Now let's go through the five steps of Method 2 using the same monthly
statement we used earlier to show the other method.

Step 1 Compute the number of days in the billing cycle. This is
done exactly as in Method 1, Step 1. The result, as before, is 31 days.

Step 2 Count the previous amount due, *including interest*, 31 times. (There are 31 days in the billing cycle.) (Notice that in Step 2 of Method 1, interest was excluded.)

$$1,291.47 \times 31 = 40,035.57$$

Step 3 (This is the step that accounts for new purchases.) Count the appropriate number of days for each purchase and multiply the amount of the new purchase by that number of days.

(a) The first purchase of $28.88 was made at the New England store on 12/07 (the transaction date). Since the closing date of the last billing cycle was 12/10, the date of this transaction is taken to be the first day of the current billing cycle, which is 12/11. Thus, this purchase is counted 31 times:

$$28.88 \times 31 = 895.28$$

(b) The purchase of $29.70 was made on 12/17. To compute the number of days, note that:

From 12/17 to 12/31 there are:	
(counting 12/17 as one of those days)	15 days
From 01/01 to 01/11 there are:	11 days
TOTAL	26 days

$$29.70 \times 26 = 772.20$$

(c) The purchase of $19.20 was made on 12/21.

From 12/21 to 12/31 there are:	
(counting 12/21 as one of those days)	11 days
From 01/01 to 01/11 there are:	11 days
TOTAL	22 days

$$19.20 \times 22 = 422.40$$

(d) The purchase of $29.16 was made on 12/27

From 12/27 to 12/31 there are:	5 days
From 01/01 to 01/11 there are:	11 days
TOTAL	16 days

$$29.16 \times 16 = 466.56$$

Step 4 Count the full payment made ($200.00) 3 times since it was "present" during only 3 days of the billing cycle

$$200.00 \times 3 = 600.00$$

Step 5 Compute the average daily balance. To do this, find the sum of the items in steps 2 and 3, subtract the result in step 4, and then divide by the number of days in the billing cycle (step 1).

$$\text{Average Daily Balance} = \frac{(\text{Step 2}) + (\text{Step 3}) - (\text{Step 4})}{\text{Step 1}}$$

$$= \frac{(40,035.57 + 895.28 + 772.20 + 422.40 + 466.56 - 600.00)}{31}$$

$$= \$1,354.58$$

Using Method 2, the average daily balance is $1,354.58, substantially different from the result we got with Method 1 ($1,253.62). The difference is $100.96.

We next have to calculate the interest charges (1.575%) on the average daily balance:

Current interest due $= 1.575\% \times \$1,354.58$
$= 0.01575 \times \$1,354.58$
$= \$21.33$

The total amount you owe the credit card company is the sum of the previous balance due, the current interest due, the total of all new purchases ($28.88 + \$29.70 + \$19.20 + \$29.16 = \$106.94$), less the amount paid:

Total due $= \$1,291.47 + \$21.33 + \$106.94 - \$200.00$
$= \$1,219.74$

According to Method 2, we owe $1,219.74 on this credit card.

The difference between the total amount due in Methods 1 and 2 is $1.59 and is the result of the way the average daily balance was computed. It reflects the difference in the amount of interest due (Method 1 $= \$19.74$; Method 2 $= \$21.33$).

Is $1.59 a significant amount of money? No one can answer that question for you. The significance of *any* sum of money depends on many factors. One factor is what it can buy. For $1.59, we can have the "breakfast special" in our neighborhood luncheonette, or buy the Sunday newspaper, three daisies, or three cups of coffee.

When it comes to credit cards, keep in mind that there will be a difference between Methods 1 and 2 *every month*. So if you kept to the identical purchasing and payment patterns, the yearly difference would be

$12 \times \$1.59 = \$19.08$ per year

If you increase the amount of your purchases, the difference would be still greater. The difference would be even more if you wait a little longer to make a payment.

Sometimes the method of computing the average daily balance is more significant than the variability in interest rates.

Here's how

SUPERCARD VS. CLASSYCARD

Let's suppose that SUPERCARD computes the average daily balance by Method 2 and charges an *annual percentage rate* of 18.00%. CLASSYCARD uses Method 1 to figure out the average daily balance, but has a 19.00% annual percentage charge. Which card offers you the most advantageous terms?

SUPERCARD

The monthly interest rate is $18\% \div 12 = 1.5000\%$.

The average daily balance on the bill used in the example in this section, computed by Method 2 = $1,354.58.

The interest charge is $0.015000 \times \$1,354.58 = \$20.32$.

CLASSYCARD

The monthly interest rate is $19\% \div 12 = 1.5833\%$.

The average daily balance (Method 1) = $1,253.62.

The interest charge is $0.015833 \times \$1,253.62 = \$19.85$.

In this example, CLASSYCARD turns out to be less expensive to you even though it charges a higher rate of interest.

Clearly, the most advantageous credit card will be the one that both charges the lowest interest and computes the average daily balance by a method that comes close to Method 1. If you are using a credit card that computes the average daily balance by Method 2, try to pay your bill as quickly as possible so that you don't pay more interest on your interest.

It is never necessary for you to go through any of the computations we showed you in this section. You can be quite certain that the credit card company's computers are doing it correctly.

However, *it is important for you to read the*

small print

Take careful note of which items are included in the average daily balance. In particular, check each of your credit card's contracts to see if they include interest *and* new purchases; if so, this is disadvantageous to you. Also, see if new purchases are being considered from the posting date (better) or transaction date (worse for you). *Try to use the credit card that gives you the best deal* even if it may mean you have to pay the entire balance each month. As we've explained before, it's all in the cards.

PART TWO

OUTDOOR MATH

5

Restaurants
and Boutiques

Section 1: Eating Out: Don't Be a Soft Touch (Estimating and Tipping)

Checking the bill and computing a tip in a restaurant is a source of anxiety for many people. While the origins of tipping are obscure, it's a practice with a long history. And, although you may sometimes want to curse the person who first gave someone a reward for performing a task, the fact is that tipping is a well-established custom that you just have to learn to cope with. Before we show you how to figure out a tip, we'll start with some general guidelines for estimating the size of your bill.

Here's how

ESTIMATING YOUR BILL

There are many times when you need to be able to approximate the amount of a bill even though you don't need to know the exact total. One such time is an unplanned trip to a store when you might like to know whether you're carrying enough money to pay for your purchases *before* you get to the checkout counter. Another time is when you're standing outside a restaurant examining the menu and trying to decide if you like the place and the selections well enough *and* if you can afford to eat there.

EXAMPLE: Let's start with a walk through the supermarket. You have $20 when you stop at the store to pick up a few items. (With today's prices, even without estimating, you already know in advance that you can't afford more than a few!) You want a container of milk (99¢), furniture polish ($2.99), 2 cucumbers (67¢), cheese ($2.19), a box of strawberries on special ($1.89), dog food (4 cans for $2.57) and a 2-liter bottle of soda ($1.29). The question is, will you have enough money to pay for all this?

✓*TIP ONE*   Whenever you estimate in these types of situations, it's better to err on the high side—that is, it's always good to overestimate. This leaves you with a cushion—and no surprises or embarrassments. So, as a general rule, *round prices up.*

✓*TIP TWO*   The second trick is that since estimating requires you to do mental addition, round the actual prices to numbers that are easy to work with, such as whole dollars, half dollars, quarters . . .

In this way:

ACTUAL PRICE		ROUNDED PRICE	CUMMULATIVE ADDITION
.99	⟶	1.00	1.00
2.99	⟶	3.00	4.00
.67	⟶	.75	4.75
2.19	⟶	2.25	7.00
1.89	⟶	2.00	9.00
2.57	⟶	2.50	11.50
1.29	⟶	1.25	$12.75, estimated
$12.59, actual			

EXAMPLE: Let's now try another example where you'll want to estimate probable costs in whole dollars. The Coral Inn Restaurant lists the following prices on the menu in the window: entrees ranging from $9.95 to $12.95; appetizers from $1.75 to $4.95; and desserts from $2.00 to $3.95. There are also salads ($1.75 to $3.00), soups ($1.50, $2.50) and beverages (from $1.00 for coffee to $1.50 for espresso). Assuming that each of you will also have a glass of wine, about how much will dinner for two cost?

To estimate, you must first try to figure out how hungry you are. Will you have a full dinner? An entree and dessert and coffee? How about salad, an entree, dessert, and coffee for one; an appetizer, entree, salad, and coffee for the other. Be generous. This is one situation where your eyes should be bigger than your stomach—so your stomach isn't bigger than your pocketbook!

Round up, use whole dollars and, again, estimate costs on the high side:

Salad, $3; + entree, $13→$16; + dessert, $3→$19; + coffee, $1→$20 (for one person); *plus* (adding in the second person):

Appetizer, $4→$24; +entree, $13→$37; +salad, $3→$40; +coffee, $1→$41.

Add the cost of the wine ($3.50/glass × 2 = $7), tax and the tip, and eating at the Coral Inn can easily amount to $55 or $60. Of course, you can probably eat for a lot less by eating a lot less—which is what some people choose to do when they want to try a restaurant slightly out of their price range.

Now that you know the basics, let's move inside the restaurant. You've had your meal, the bill arrives, and you're faced with an important choice: whether or not to estimate the accuracy of the bill or to calculate it exactly. We frequently estimate, but that's only because *we are willing to be wrong* by a small amount, especially on a large check. By estimating the accuracy of the total bill rather than checking it exactly, we are, in effect, saying that we don't care if there's a small error in the other party's favor—that we are willing to pay as much as a couple of dollars extra for the *luxury* of not adding up the bill.

Here's what we do when we want to know if a check is *approximately* correct. First, we make sure that only those items we ordered (and received) are included on the bill. Second, we check to see that the charge for each item is the same as the charge listed on the menu (or on the price tag). In most restaurants these days, the arithmetic part (the adding and the computation of taxes) is done by computerized cash register, so most errors are inadvertent charges for items you didn't order or ordered and never received. It is also common for the price of the item to be wrong. Therefore, whether you're estimating the accuracy of the total bill or calculating it exactly, always go through these two steps.

Now you're ready to round the dollar amounts to numbers that are easy to add in your head—whole dollars, halves, quarters—and add up the items cumulatively as you go down the column of figures. On a $30 bill, your estimated total and actual total should agree within a couple of dollars.

Going down the check you come to a subtotal, then to the tax. To understand how the sales tax (and tip) is computed, you'll need to understand percentages.

Here's how

COMPUTING THE TAX EXACTLY

In many places in the United States there is a state and/or city sales tax on restaurant meals; maybe on the nonfood items in your grocery basket as well, such as furniture polish, soap, and so on; and perhaps on other purchases like clothes or appliances. These taxes can range from as low as 1% or 2% to as high as the 8.50% in some counties in New York state. Taxes are generally applied to your total restaurant bill: food and bar beverages.

Calculating the amount of sales tax involves two computations:

> **Step 1** Convert the sales tax percentage to a decimal by dividing it by 100. (To divide by 100, move the decimal point 2 places to the left.)
>
> **Step 2** Multiply the price of the item(s) by this decimal. The result is the sales tax.

That's all there is to it, although you might want to find the total cost, including tax, by adding the sales tax to the original price.

EXAMPLE 1: Dinner for two comes to $43.75. Local state and city sales tax is 6%. Find the dollar amount of the tax and the total cost of the dinner, including tax.
SOLUTION:
  (1) Convert 6% to a decimal:
      6% = 6/100 = 0.06
  (2) Multiply the price by the decimal to find the tax:
      0.06 × $43.75 = $2.63 = Tax (rounded to two decimal places).
  (3) Add the tax to the price of the dinner to find the total cost:
      $2.63 + $43.75 = $46.38 = Total Cost

EXAMPLE 2: In New York City the sales tax is 8¼%. If a new car lists for $12,870, how much will the car cost us and how much of this will we by paying in tax?
SOLUTION: Step 1 involves converting the sales tax percentage to a decimal. Converting 8¼% requires first converting ¼ to a decimal. If you don't recognize that ¼ = 0.25, you can figure this out by dividing 1 by 4. (All fractions can be converted to decimals by dividing the numerator, the number on the top of the fraction, by the denominator, the number on the bottom.) Thus, 8¼% = 8.25%. Now:
      8.25% = 8.25 ÷ 100 = 0.0825
      Tax = 0.0825 × $12,870 = $1,061.78
      Total Cost = $1,061.78 + $12,870 = $13,931.78

With paper and pencil, or better yet with a pocket calculator, it's always possible to figure out the tax *exactly,* but in a restaurant you may be satisfied with an approximation.

ESTIMATING THE TAX

Approximating the tax involves mental arithmetic and three steps:

Step 1 Round off the bill to the *nearest* whole dollar.

Step 2 Multiply the whole dollar amount by the tax, forgetting about the decimal. (Multiply by 4 for 4%, 3 for 3%, 8 for 8¼%, etc.)

Step 3 Now consider the decimal and in your answer move the decimal point two places to the left.

EXAMPLE: What is the approximate tax on $17.27 if the tax rate is 8.25%?
SOLUTION:
(1) $17.27→17
(2) 17×8 (we're simplifying 8.25) = 136
(3) 136/100 = 1.36

The approximate tax on $17.27 is $1.36, which is pretty close to the exact tax of $1.42. We're slightly off because we rounded (down) both the dollar total and the tax rate so that we would have numbers that were easier to work with.

It may happen that the only number that appears on a bill is the *total* price, including tax. (Let's call this the *gross price*.) In such cases, you may want to figure out the *net price* (the price before the tax was added) and the amount of the tax itself. This kind of problem commonly occurs in retail businesses, for example, when the cashier fails to list the tax separately. Knowing only the gross receipts, the manager/owner must determine the net receipts and tax in order to know how much he owes to the local government.

Here's how

COMPUTING NET PRICE AND TAX

To find the amount of tax when you know only the tax rate and gross price (and not the net price) requires using two formulas. First, to compute the net price, substitute:

$$\text{Net price (receipts)} = \frac{\text{Gross price (receipts)}}{1 + (P/100)}$$

where P is the tax percentage.*

Then to find the amount of tax, subtract the net price from the gross price as follows:

$$\text{Tax} = \text{Gross price (receipts)} - \text{Net price (receipts)}$$

As an illustration, suppose that a restaurateur finds that the gross receipts for December 12 totaled \$876.24. She needs to determine how much of that amount is sales tax, knowing that the local tax rate is 5%.

To solve this problem, calculate the net receipts according to the first formula given above:

$$\text{Net receipts} = \frac{\text{Gross receipts}}{1 + (P/100)}$$

$$= \frac{\$876.24}{1 + 5/100}$$

$$= \frac{\$876.24}{1 + .05}$$

$$= \frac{\$876.24}{1.05}$$

$$= \$834.51$$

Now, to find the tax:

$$\text{Tax} = \text{Gross receipts} - \text{Net receipts}$$
$$= \$876.24 - \$834.51$$
$$= \$41.73$$

The only thing we haven't considered is the tip and how it's computed. *Here's how*

CALCULATING THE TIP

---

*N = Net price; G = Gross price
N + (P/100)N = G
[1 + (P/100)]N = G
N = G/[1 + (P/100)]

A tip of whatever size you care to leave should be computed on the amount of the check *before* the tax is added on.

There are exact ways to calculate a tip as well as estimates that you might care to use. Estimating the tip, like estimating anything else, involves some deviation above or below the exact amount, but you can easily compensate for this by subtracting or adding a little.

Except in certain situations like coat checks and baggage handling where tips are usually figured on a per piece basis, tips are expressed in percentages. These percentages must be converted to decimals in the same way as we did with the sales tax. Let's consider a tip of 15% which has become an accepted standard for good food and good and amicable service. Remember to convert the percentage to a decimal:

$$15\% = \frac{15}{100} = 0.15$$

Suppose your check came to $36.72 before tax. To calculate a 15% tip *exactly:*

$$0.15 \times 36.72 = 5.51$$

✓*TIP*   Computing 15% of an amount like $36.72 is pretty difficult to do in your head. To make it easier, remember that:

$$15\% = 10\% + 5\%$$

So an easy way to figure out what 15% comes to is to calculate 10% and then add to it half of that amount (5%).

Here's how
First, compute 10% of 36.72:

$10\% \times 36.72 = 3.672$ or 3.67 (Multiplying by 10% is the same as moving the decimal point one place to the left)

Since 10% is 3.67, half of that is about 1.84. So the 15% tip is:

$$3.67 + 1.84 = 5.51$$

This is precisely what we got when we computed it by multiplying by 0.15.

To calculate 15% of $36.72 in your head, you might want to round off the dollar amount to $37. Then 10% of $37 = $3.70 + half of that is $1.85. Then add, $3.70 + $1.85 = $5.55, which is a little higher than the exact calculation because we rounded up.

Let's do another example both ways. What is a 15% tip on a check of $27.75?

SOLUTION (exact):
  (1) Calculate 10% of 27.75   = 2.78
  (2) Take ½ of that amount (5%) = 1.39
  (3) Add the 10% and the 5%   = 4.17

SOLUTION: (approximate):
  (1) Round $27.75 to $28
  (2) Calculate 10% of 28   = 2.80
  (3) Take ½ of that amount   = 1.40
  (4) Add the 10% to the 5%   = 4.20

✓✓*HOT TIP*   There are other quick tricks to use to approximate a 15% tip. In New York City the sales tax is 8.25%, so just double the tax. By doubling the amount of the tax, you get 16.50%, which is pretty close to 15%. So if you want to leave about 15%, just subtract a little bit.

You can also do this if the tax where you live is 7%. Double it and you have 14%. Add a little to the total to come close to 15%. If your sales tax is 5%, multiply the tax shown by 3 to get 15%. Similarly, if your sales tax is 3%, multiplying the amount shown by 5 will result in 15%.

Tips were once an expression of appreciation for *extra* services. Today, tips are expected, and many jobs are dependent on them. A 15% tip has become the unwritten standard for reasonable service, while 20% rewards outstanding service. A small tip, say 5%, or no tip at all is your way of saying you were dissatisfied. In that case, however, it's doubly effective if you also describe the problems you encountered to the waiter or manager.

While we may believe that the quality of service and social custom actually determine the size of the tip, an article in the April 4, 1985 edition of the New York *Daily News* makes it clear that other factors can also play a part. The *News* reported that a psychology professor carried out a study on ways to increase the tip. In the study as the customers were given their change, they were either touched lightly on the hand, touched on the shoulder, or not touched at all. Customers who were touched on the shoulder tipped an average of 14.4%; those who were touched on the hand tipped 16.7%; and those who weren't touched at all tipped 12.2%, on average. We think the moral might be "don't be a soft touch!"

Section 2: What a Buy! Discounts and Markups

Since we are redecorating, we were particularly interested in the following ad that appeared in this morning's paper—

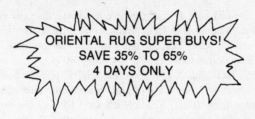

—which illustrates one of the problems of discounting, that of determining the price you pay for an item after the discount is taken. This is the sale or discounted price and we'll show you two ways to compute it.

Method 1

This method is for situations where you know the original price of the item and the percent discount, and you want to find the sale price. It allows you to figure out both how much money you'll be saving (the dollar amount of the discount) *and* the price of the item after the discount is subtracted. This method involves three steps:

Step 1 First convert the percent discount to a decimal by dividing by 100. (This is the same as moving the decimal point 2 places to the left.)

Step 2 Then multiply the original price by the decimal number found in Step 1. This is the amount of money you save—the discounted amount.

Step 3 Finally, subtract the amount of the discount from the original price to find the sale price.

Let's do an example. Suppose the original price of the oriental rug was $1,450 and it's on sale at 35% off. To find the sale price:

(1) $35\% = 35/100 = 0.35$
(2) 35% of $\$1,450 = 0.35 \times \$1,450 = \$507.50$
(3) $\$1,450 - \$507.50 = \$942.50$

In this example, the discount comes to $507.50 (the amount in Step 2) and the rug on sale costs $942.50 (Step 3).

There is another way to compute the sale price that is more direct, saves some work, and eliminates one source of arithmetic error. To understand it requires a little more insight into the discounting process.

Method 2

We just showed how to find the amount of the discount (the amount saved) and then subtracted it from the original price to arrive at the sale price. However, it's really not necessary to calculate the discount in order to figure out the sale price.

Follow this reasoning. Think about the original price as being 100% of itself. The discount is 35% of the original price. After subtracting the discount percentage from 100%, you are left with 65% of the original price (100% − 35% = 65%). This remainder is the sale price.

Therefore, to find the sale price using Method 2:

> **Step 1** Subtract the discount percentage from 100%. Let's call the answer the "sale price percentage" (in this case, 65%).
> **Step 2** Convert the sale price percentage to a decimal by dividing by 100; move the decimal point 2 places to the left.
> **Step 3** Multiply the original price by the decimal (from Step 2). The result is the sale price.

Let's use the same illustration of the rug that cost $1,450 originally and that is on sale at 35% off. Again, we want to find the sale price. To solve the problem using Method 2:

> (1) 100% − 35% = 65%
> (2) 65% = 65/100 = 0.65
> (3) 0.65 × $1,450 = $942.50

✓*TIP*   Method 2 is very useful when you're in a store trying to figure out if you want to buy an item on sale: it gives you the sale price quickly and in one step. As we would hope, Methods 1 and 2 give the same result. But this second method has several advantages in addition to speed. By eliminating the need to subtract the calculated discount amount from the original price (Step 3, Method 1), we also reduce the chance to making an error. Method 2 is also easier to do in your head because it's a lot easier to do just one mental multiplication (Method 2) than it is to do both a multiplication and a subtraction (Method 1).

You can also use this method for a quick approximation by rounding both the sale price percentage and the original price to numbers that are easy to multiply in your head: We round 0.65 up to 0.70 and round $1,450 down to $1,400. Multiplying 0.7 by $1,400 we find an approximate sale price of $980, which is a bit higher than the actual sale price.

We're now ready to turn to the second problem associated with discounting, which is how to compute the *percent discount*.

Here's how

COMPUTING THE % DISCOUNT

Another ad for an oriental rug quoted the sale price of a 6 foot by 9 foot rust and blue antique rug as $999. The ad also described its value as $2,500, making the sale price sound like a really good bargain. How good a buy becomes more clear when we figure out the percent discount—the percentage we are getting off. Again, this involves three steps, as follows:

Step 1 Find the difference between the original price and the sale price.

Step 2 Divide the difference by the original price.

Step 3 Convert the result of Step 2 to a percent by multiplying by 100.

These operations can be summarized by a formula:

$$\text{Percent discount} = \frac{\text{Original price} - \text{Sale price}}{\text{Original price}} \times 100\%$$

Let's do an example from the ad that gave the rug's sale price as $999 and the original value as $2,500. What we want to do is to find the percent discount, assuming that the quoted value was the original price. Substituting in the formula:

$$\text{Percent discount} = \frac{\$2,500 - \$999}{\$2,500} \times 100\%$$

$$= \frac{\$1,501}{\$2,500} \times 100\%$$

$$= 0.6004 \times 100\%$$
$$= 60.04\% \text{ or } 60\%$$

Sixty percent is really an excellent discount—*if* the rug was really worth $2,500. Unfortunately, many such supposed "values" or list prices are often inflated so that the discount will sound bigger. This makes people feel like they're getting a better bargain.

The opposite of discounts are *markups*. Business people "mark up" the cost of an item over what they paid for it to arrive at the price they will charge. The markup is also known as the *margin* (the retail food business operates on a "small margin") or *gross profit,* which is the "profit" before costs, including overhead, are subtracted. The relationship between cost, markup, and retail or selling price can be summarized this way:

Cost + Markup = Selling price

A markup is quoted as either a percentage of the dealer's cost or of the selling price of the item depending on accepted practices of the particular type of business or industry. We'll show you how both types of markups are computed.

Here's how

Method 1

There are two methods for computing the selling price when the markup is based on the cost of the item. They are very similar to the ways we showed you earlier of calculating the sale price when the percent discount is known. As was the case there, Method 1 involves 3 steps:

Step 1 Convert the markup percentage to a decimal by dividing by 100. (Move the decimal point 2 places to the left.)

Step 2 Multiply the cost of the item by this decimal to find the dollar amount of the markup.

Step 3 Add the markup dollar amount to the cost of the item to obtain the selling price.

EXAMPLE: Toodles department store buys one make of suit for $225. The store marks the suit up 80% based on its cost. Find the selling price.

SOLUTION:

(1) $80\% = 80/100 = 0.80$

(2) Markup $= 0.80 \times \$225 = \$180$

(3) Selling price $= \$225 + \$180 = \$405$

The other way to compute the selling price when the percent markup is based on known cost avoids the last step of addition.

Method 2

This method considers the cost of the item to the store as 100% of itself. In this case, follow the 3 steps below to calculate the selling price.

Step 1 Add the given percent markup to 100%.

Step 2 Convert the total percent obtained in Step 1 to a decimal by dividing by 100.

Step 3 Multiply the cost of the item by the decimal (from Step 2). The result is the selling price.

Let's redo the same example that we used to illustrate the first method where a suit cost a department store $225, and the store's markup is 80% of cost. To find the selling price using Method 2:

(1) 100% + 80% = 180%

(2) 180% = 180/100 = 1.8

(3) Selling price = 1.8 × $225 = $405

✓*TIP*   By eliminating the last step of Method 1 (which requires adding the amount of the markup to the cost to arrive at the selling price), Method 2 enables you to determine the selling price faster and in a way that has less possibility for arithmetic errors.

How can we compute the selling price when the markup is based on the selling price? At first glance it may seem impossible to do, except maybe by guessing. However, while it is not impossible, it does require some fairly involved algebra. Here instead, we'll leave out the algebraic manipulations and just show you how to do the computations. If it seems mysterious, remember that we're only showing you the end result.

Here's how

% MARKUP BASED ON SELLING PRICE

When the markup percentage (P) is known to be based on the selling price, we can figure out the selling price according the the following formula—provided we also know the cost*:

---

*(C = Cost; S = Selling Price)

$C + (P/100)S = S$

$C = S[1 - (P/100)]$

$S = C/[1 - (P/100)]$

$$\text{Selling price} = \frac{\text{Cost}}{1 - (P/100)}$$

Suppose the suit in the previous example still costs Toodles $225 but the store's markup is computed as 80% of the *selling price*. To find the selling price, substitute in the formula:

$$\text{Selling price} = \frac{\$225}{1 - (80/100)}$$

Then carry through the calculations:

$$= \frac{\$225}{1 - 0.80}$$
$$= \frac{\$225}{0.20}$$
$$= \$1,125$$

That's quite a price for a suit! We didn't make a mistake, but rather chose this example to illustrate the effect of computing markups as a percentage of selling price. For any given percent markup, the markup is always higher when it is computed on the basis of selling price than when it is figured as a percent of cost. This is because the selling price is always higher than the cost. The best way to find out whether the markup is based on cost or on selling price is to ask!

Markups based on selling price are not generally used in the clothing industry, but are typical, for example, of the way cosmetics are priced.

If both the selling price and the cost of an item are known, you can compute the markup percentage that is based on cost.

Here's how

FINDING THE MARKUP % BASED ON COST

There's a formula for use in situations in which you want to find the markup percentage and you know it's based on cost; you also know the cost and the selling price. This is the formula:

$$\% \text{ Markup} = \frac{\text{Selling price} - \text{Cost}}{\text{Cost}} \times 100\%$$

EXAMPLE: If a suit is selling for $470 and it cost the store $290, what is the percent markup based on cost?

SOLUTION:

$$\% \text{ Markup} = \frac{\$470 - \$290}{\$290} \times 100\%$$
$$= \frac{\$180}{\$290} \times 100\%$$
$$= 0.62 \times 100\%$$
$$= 62\%$$

If the selling price and cost are known, you can also compute the markup percentage that is based on *selling* price.

Here's how

FINDING THE MARKUP % BASED ON SELLING PRICE

If you know that the markup percentage is based on selling price, and if you also know the selling price and the cost, use this formula to find the markup percent:

$$\% \text{ Markup} = \frac{\text{Selling price} - \text{Cost}}{\text{Selling price}} \times 100\%$$

(Please take note of the fact that the only difference in the formulas for computing percent markup based on selling price and percent markup based on cost is that, in the former case, you divide by the *selling price* while in the latter case, you divide by the *cost*.)

Let's find the percent markup based on the selling price of a suit selling for $470 that cost Toodles department store $290.

SOLUTION:

$$\% \text{ Markup} = \frac{\$470 - \$290}{\$470} \times 100\%$$
$$= \frac{\$180}{\$470} \times 100\%$$
$$= 0.383 \times 100\%$$
$$= 38.3\% \text{ or about } 38\%$$

Notice that the same dollar markup of $180 is a smaller percentage of the selling price than it is of the cost price because the selling price is higher than the cost.

Now let's find the *cost* when the percent markup is based on cost.
Here's how

FINDING COST: % MARKUP BASED ON COST

If you happen to know the percent markup (P) and the selling price, you can always figure out the dealer's cost. This is the formula*:

$$\text{Cost} = \frac{\text{Selling price}}{1 + (P/100)}$$

A VCR sells for $325 at Mad Louie's. It's well-known that Mad Louie operates on an 8% markup. What is Mad Louie's cost?

To answer this question, substitute in the formula and carry through the calculations:

$$\begin{aligned}
\text{Cost} &= \frac{\$325}{1 + (8/100)} \\
&= \frac{\$325}{1 + .08} \\
&= \frac{\$325}{1.08} \\
&= \$300.93
\end{aligned}$$

This VCR costs Mad Louie $300.93. You can always find the dealer's cost if you know the selling price and the percent markup.

You can also find the dealer's cost when the markup is based on the selling price.
Here's how

FINDING COST: % MARKUP BASED ON SELLING PRICE

---

* (S = Selling price; C = Cost)

  C + (P/100)C = S

  C[1 + (P/100)] = S

  C = S/[1 + (P/100)]

In this type of markup situation, you can calculate the cost to the dealer by finding the dollar amount of the given markup percentage of the selling price (P) and subtracting that from the selling price. (You can also do this computation in one step.) The simple formula is as follows:

$$Cost = [1 - (P \div 100)] \times Selling\ price$$

As an example, at Friendly Phyllis's, cosmetics are generally marked up 40% based on selling price. If Rose Creme lotion sells for $8.99, what is Friendly Phyllis's cost?

SOLUTION:
$$
\begin{aligned}
Cost &= [1 - (40/100)] \times \$8.99 \\
&= (1 - 0.40) \times \$8.99 \\
&= 0.60 \times \$8.99 \\
&= \$5.39
\end{aligned}
$$

Friendly Phyllis buys a bottle of Rose lotion for $5.39.

Understanding how discounts are computed lets consumers assess how much of a savings they can realize from sale items.

✓*TIP*   You can almost always save more if you buy from reputable dealers and wait for seasonal sales, such as summer rug, furniture, and bedding sales, and the mid-February sale on coats. Similarly, fruits and vegetables are usually cheaper "in season," despite the fact that many are now available year round because of advances in freezing, storage, and transportation.

Being able to figure out markups, costs, and selling prices gives you an advantage in negotiating realistically since you have a feel for the dealers' costs. Familiarity with the concepts of discounts and markups, and facility in computing them, increases your power as a consumer.

6

Foreign Travel

Section 1: An American Abroad: Dromedaries, Drachmas, and Dollars (Currency Conversion)

So you're going to travel out of the country . . .

Once you've plotted your itinerary and planned your wardrobe, the remaining nagging worry is coping with the foreign country's currency—those multi-colored, over- and under-sized bills and funny-shaped coins that will buy an expensive memento—or cheap souvenir—if you could only figure it out!

Unless you can translate foreign currency amounts into your own currency—American dollars—you'll be playing with monopoly money, never certain whether you've paid for "Boardwalk" or "Mediterranean Avenue."

In dealing with foreign currency, you should first decide whether you want to know *exactly* how much you are going to spend in American dollars or whether a *reasonable dollar approximation* will suffice. Since approximation always involves some error—up or down—your budget and the price tag are often the determining factors.

For many purchases, small ones or extremely costly ones where being a few dollars off won't make that much difference, most travelers are generally content with knowing the approximate dollar amount. In choosing a restaurant, for example, it's important to know if dinner will cost $8 or $80. But it may not matter if you estimate and figure on $6 when the bill will actually come to $7.50 or figure on $83 when the meal will end up costing $90.

However, *if the exact amount matters* (and toward the end of a trip, the dollars and cents you gained or lost because of estimating do have a way of adding up), you can always do an exact conversion—by hand with paper and pencil—or, preferably, by calculator.

Your pocket calculator is the most precise, efficient, and fastest way of converting drachmas into dollars. Small calculators are so much in evidence in the streets, at market stalls, and in shops and restaurants that using one won't label you any more "foreign" than a camera, walking shoes, or large carry-all.

Whether you decide that you want to be exact or approximate, however, your goal is the same: to determine the price in dollars. To do this, *you need to know the value in dollars of one unit of the foreign currency*. In other words, how much is *one* drachma (or one franc, one mark, one pound, or one lira) worth in American dollars?

Here's how

FINDING THE EXACT VALUE IN DOLLARS OF ONE UNIT OF FOREIGN CURRENCY

1 drachma = ?$

You can find the exact value in dollars of a unit of foreign currency by looking the rate up in a newspaper or checking at a bank or exchange bureau for the current exchange rate.

Foreign exchange rates, which change every banking day, appear daily in the business section of *The New York Times* and other major newspapers under "Foreign Exchange."

On August 12, the exchange rate was:
one Italian lira = $0.000773
one French franc = $0.17
one British pound = $1.70

Some U.S. listings only express the exchange rate as the number of units of foreign currency you can get for one dollar. These listings would look like this:

$1 = 1293.50 Italian lire
$1 = 5.88 French francs
$1 = .59 British pounds

In this case, you must convert to get the value in dollars of one unit of foreign currency.

The value in $ of one unit of foreign currency $= 1 \div$ The number of units of foreign currency per dollar

So, on August 12:
one Italian lira $= 1 \div 1293.50 = \$0.00077$ (rounded off)
one French franc $= 1 \div 5.88 = \$0.17$
one British pound $= 1 \div .59 = \$1.69$

By calculator (for lire):
PRESS 1 ÷ 1293.50 =

Once you have calculated the value in dollars of one unit of foreign currency (irrespective of whether you looked it up or computed it), you are now ready to find the dollar amount of *any number* of units of the foreign currency. It is at this point that you can either estimate or do an exact conversion as the particular situation arises. Let's do it exactly first.
Here's how

FINDING THE EXACT DOLLAR VALUE OF ANY NUMBER OF UNITS OF FOREIGN CURRENCY

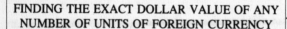

298 francs = ?$

Let's say you're shopping in Paris and you find a shirt that costs 298 French francs. To determine whether you want to buy it, you need to know its equivalent value in dollars.

Remember, in our example, the French franc $= \$0.17$

The $ value of the foreign currency $=$ The $ value of one unit of the foreign currency $\times$ The number of units of the foreign currency

So, the dollar value of:
298 francs $= \$0.17 \times 298 = \$50.66$
54,850 lire $= \$.00077 \times 54,850 = \$42.23$
46 pounds $= \$1.69 \times 46 = \$77.74$

By calculator (for francs):
PRESS 0.17 × 298 =

There may be times when you need to reverse the process: to convert dollars into foreign currency amounts.
Here's how

| FINDING THE EXACT FOREIGN CURRENCY VALUE OF ANY NUMBER OF AMERICAN DOLLARS | $500 = ?francs |

As a tourist you will need this conversion most often when you are buying foreign currency and you want to know how many of "them" you'll get for a specified number of dollars. How many francs (or pounds or lire) will you get for $500?

| The foreign currency value of a given number of dollars | = | The number of dollars | ÷ | Value in dollars of one unit of foreign currency |

Therefore:

$500 ÷ $1.69 per pound = 295.86 pounds
$500 ÷ $0.17 per franc = 2,941.18 francs
$500 ÷ $0.00077 per lira = 649,350.64 lire

By calculator (for pounds):
PRESS 500 ÷ 1.69 =

This is the same type of conversion the bank or exchange bureau does when selling you foreign currency. The actual amount you receive, however, is always somewhat less than you have computed because they charge a commission (fee) for the transaction. The fee is equal to the difference between the amount of money you get and the amount you figured you would receive, according to the conversion formula.

We're now ready to estimate.

Here's how

| ESTIMATING THE DOLLAR VALUE OF FOREIGN CURRENCY | drachmas = ? about$ |

You may well ask, why bother estimating when the exact conversion of foreign currency is so streamlined by the calculator? Take the dollar value of one franc, multiply by the number of francs and, *voilá*, you have a precise dollar and cents answer.

But sometimes it is necessary to estimate: you forgot your calculator, the battery died, you left your reading glasses on the night table and can't see the display.

Also, some people think it's fun to approximate the dollar cost of items. It gives them a sense not only of mastery, but also of belonging—of really being in a foreign place. They feel international and have a good feeling about themselves when they can approximate speedily, while achieving reasonable accuracy.

Estimating is done in your head and involves little tricks, such as rounding numbers so that they are easier to work with. The more you practice, the easier it becomes to think of little shortcuts that simplify the mental arithmetic for you. But remember that even the best schemes always involve an error factor—the dollar amounts you arrive at are, at best, close. They are rarely exact. So, every time you want to convert foreign currency into American dollars, you should first decide whether you want a close or a precise answer. (Exact numbers require a calculator.)

To do either, however, you must know the value in dollars of one unit of foreign currency. *Find this out before you start your trip.*

✓*TIP*   The strategy we are presenting for estimating the dollar value of any number of units of foreign currency can be applied to any and all of the world's currencies: that's why the conversion table on page 126 looks complicated. It, and the other steps, cover *all* currencies. However, you're only interested in one currency conversion at a time, so all you have to do is:

- Pick the case that describes the exchange rate of the currency you're interested in (Cases B and C require some preliminary steps).
- Find the line in the conversion table on page 126 that shows the shortcut multiplications and/or divisions.
- Do the one or two steps described in the conversion table in your head.

Don't be put off by the steps and the different cases. Remember, you're only going to be dealing with one country's currency at a time that will involve only one case and one set of short steps. If you do a few practice examples with that currency, the approximate conversion will soon become second nature.

Ready?
Here's how

SELECTING THE CASE THAT FITS THE DOLLAR VALUE OF ONE UNIT OF FOREIGN CURRENCY

You know the dollar value of one unit of the currency of the country you're interested in: one Italian lira = $0.00077, one French franc = $0.17, one English pound = $1.69.

First, select the case that best describes this dollar value.

Case A The $ value of a unit of foreign currency is between $0.10 and $0.99 (example, one French franc).

Case B The $ value of a unit of foreign currency is less than $0.10 (this covers any amount that has one or more zeros to the right of the decimal; example, one Italian lira).

Case C The dollar value of a unit of foreign currency is $1 or more (example, one British pound).

(Remember, you only do this one time for each different foreign currency.) Now, skip to the appropriate case.

CASE A: ESTIMATING THE DOLLAR AMOUNT OF FOREIGN CURRENCY WITH A $ VALUE PER UNIT OF BETWEEN $0.10 AND $0.99

This case describes most European currencies and is also the one involving the fewest steps.

A shirt costs 278 francs; what is its approximate dollar amount?

Step 1 Find the number in the conversion chart that is closest to the dollar value of one franc. Given that one franc = $0.17, the closest number is the same, 0.17. (Note that for other currencies the exact number probably won't appear in the table, so use the *closest* number.)

Step 2 Follow the directions for 0.17 in column III of the chart, which says, "Divide by 6." 278 ÷ 6 is about $47 (since 278 is about 280, you can also do 280 ÷ 6, or $47).

For comparison, the exact value is 0.17 × 278 = $47.26.

SAMPLE PROBLEM: Given that one Canadian dollar = $0.87, convert 35 Canadian dollars to American dollars.

SOLUTION: In Table 1, 0.87 is closest to 0.90. The shortcut direction is to first multiply by 9, then move the decimal point one place to the left.

Multiplying 35 × 9 gives 315 and moving the decimal point one place to the left gives $31.5 or $32. For comparison, the exact value is 0.87 × 35 = $30.45.

Table 1
Conversion Table Directions

I CLOSEST DECIMAL NUMBER	II APPROXIMATE (OR EXACT) FRACTIONAL EQUIVALENT	III DIRECTIONS FOR SHORT-CUT MULTIPLICATION
0.10	$\dfrac{1}{10}$	Move the decimal one place to the left
0.13	$\dfrac{1}{8}$	Divide by 8
0.17	$\dfrac{1}{6}$	Divide by 6
0.20	$\dfrac{2}{10}=\dfrac{1}{5}$	Multiply by 2 and move the decimal point one place to the left
0.25	$\dfrac{1}{4}$	Divide by 4
0.30	$\dfrac{3}{10}$	Multiply by 3 and move the decimal point one place to the left
0.33	$\dfrac{1}{3}$	Divide by 3
0.40	$\dfrac{4}{10}=\dfrac{2}{5}$	Multiply by 4 and move the decimal point one place to the left
0.50	$\dfrac{1}{2}$	Divide by 2
0.60	$\dfrac{6}{10}=\dfrac{3}{5}$	Multiply by 6 and move the decimal point one place to the left
0.67	$\dfrac{2}{3}$	Divide by 3, then multiply by 2
0.70	$\dfrac{7}{10}$	Multiply by 7, then move the decimal point one place to the left
0.75	$\dfrac{3}{4}$	Divide by 4, then multiply by 3
0.80	$\dfrac{8}{10}=\dfrac{4}{5}$	Multiply by 8 and move the decimal point one place to the left
0.90	$\dfrac{9}{10}$	Multiply by 9 and move the decimal point one place to the left
0.95+	1	Nothing to do (just multiply by 1)

CASE B: ESTIMATING THE DOLLAR AMOUNT OF FOREIGN CURRENCY WITH A $ VALUE PER UNIT OF LESS THAN $0.10

When the dollar value of a unit of foreign currency is less than $0.10, there are more than 10 units to the dollar. (We saw earlier that there were actually 1,294 lire to a dollar.) In this situation, the shortcut conversion requires getting rid of zeros (by moving the decimal point) so as to make the numbers you are working with smaller (so you can do the arithmetic in your head).

Here's how

A sweater costs 27,500 lire; approximately how much is this in dollars?

Step 1 First, look up the dollar value of a unit of foreign currency. Then, to find your new number, move the decimal point (to the right) until it is just to the left of the first non-zero digit. (Count how many places you moved.) Now you can find the number in the table that is closest to your new number.

Given that one lira = $0.00077, moving the decimal point 3 places to the right gives you 0.77.

Step 2 To convert the total amount of foreign currency on the price tag, move the decimal point in the price as many places to the *left* as you moved it to the right in Step 1.

Thus 27,500 lire becomes 27.500 or about 28. (If by doing this you have no numbers remaining before the decimal, the item costs less than $1. If the item was tagged at 275 lire, for example, moving the decimal 3 places to the left would give 0.275—which is $0.21, much less than $1.)

Step 3 Follow the directions in Table 1 for the number you looked up, applying them to the foreign currency amount remaining after you moved the decimal place.

In our example, 0.77 is closest to 0.75. The directions in column III for this number are to "Divide by 4, then multiply by 3." (So 28, divided by 4 is 7, and multiplying by 3 gives you $21.)

For comparison, the exact value is $0.00077 \times 27,500 = \$21.18$.

Did this seem a little complicated? Don't worry, it will get easier with just a little practice because Step 1 is only done one time for each different currency.

✓*TIP* Do Step 1 before you set off on your trip to Italy. Also, learn the directions for this number and do 3 or 4 practice examples. You'll see that it becomes almost automatic and can be computed very quickly.

SAMPLE PROBLEM: Convert 5,620 lire to dollars.

SOLUTION: We have already done Step 1 for lire at $0.00077, moving the decimal 3 places right and finding a number close to 0.77 on the conversion table. Now, taking the 5,620 lire, we move the decimal point 3 places to the left to get 5.620. Approximate 5.620 to 6, then divide by 4 (to obtain 1.5), and multiply by 3 to get a corrected price of $4.50.

For comparison, the exact value is $0.00077 \times 5,620 = \$4.33$.

CASE C: ESTIMATING THE DOLLAR AMOUNT OF FOREIGN CURRENCY WITH A $ VALUE PER UNIT OF $1 OR MORE

Currencies that fit this situation include British pounds (at the present rate of exchange) and currencies from some other Commonwealth nations. In *all* Case C instances, your answer will be a "larger" number than appears on the price tag.

The shortcut conversion has 3 steps.

A hat costs 15 British pounds. About how many dollars is this?

Step 1 Multiply the price tag amount by the *dollar portion* of the dollar value of one unit of foreign currency.

The dollar value of one pound is $1.69. Of this amount, "1" is the dollar portion. Thus, we multiply $15 \times 1 = 15$.

Step 2 Find the number in the conversion table closest to the *decimal* portion of the dollar value and follow the directions in Column III.

The closest number to 0.69 is 0.70. The directions for .70 tell you to "multiply by 7, then move the decimal point one place to the left." That is, $15 \times 7 = 105$ or 10.5 after you move the decimal point.

Step 3 Add the resulting numbers from Steps 1 and 2 together.

In our example this would be: $15 + 10.50 = \$25.50$.

For comparison, the exact value is $\$1.69 \times 15 = \$25.35$

SAMPLE PROBLEM: Given that one punt (Ireland) = $1.54, convert 25 punts to American dollars.

SOLUTION: Multiply 25 by 1 (25×1) to give 25. Look up 0.54 in Table 1 (the closest number is 0.50) and "divide by 2" ($25 \div 2 = 12.50$), and "add the two numbers together" ($25 + 12.50 = \$37.50$).

For comparison, the exact value is $1.54 \times 25 = \$38.50$.

If you read through all three cases *without* working through each step, you will probably be thoroughly confused by now. To convince yourself that estimating is fast and fairly accurate, you need to work through the steps of the problem that applies to your travels. Do so now to develop some facility with these procedures.

✓✓*HOT TIP* Keep in mind that much of the "work" involved in estimating is done once and only once for each foreign currency and can be done *before* you even board the plane.

Wherever you're going, find out in advance the dollar value of one unit of that country's currency. This piece of information tells you whether you have a Case A, B, or C conversion situation; it also tells you the arithmetic directions for the shortcut conversion.

So, if you're going to France, leave the U.S. knowing that the franc is $0.17. This is a Case A currency situation, so all you have to do is "divide by 6" (the directions in the conversion chart for 0.17). In Paris and on the Riviera, then, all prices are divided by 6 to give you the approximate dollar value.

If you're off to Italy (where the lira = $0.00077, a Case B conversion), you have memorized moving the decimal 3 places to the right and the directions for 0.75. To estimate the dollar amount of an item, what you do is first move the decimal 3 places to the left (to give you a handier number to work with), divide by 4 and, finally, multiply by 3.

Before you start out for London, you've learned the dollar value of one British pound and the conversion directions for the number nearest the decimal portion of this value. If the pound = $1.69, estimating the dollar value then entails only "multiplying by 7, then moving the decimal point one place to the left" and adding this answer to the number of pounds (since pounds were multiplied by one).

That's really all there is to it.

Section 2: Going International: Temperature Conversion

If you've grown up in the United States and have done little traveling abroad, you've probably not given much thought to the fact that temperature is measured in degrees *Fahrenheit (°F)* in this country and degrees *Celsius (°C)* in most other places. The Fahrenheit scale was named after the German physicist, Gabriel Daniel Fahrenheit (1686–1736), who devised it and also improved the thermometer by substituting mercury for other materials.

Like other temperature scales (such as Celsius, Kelvin, or absolute), the Fahrenheit scale has two fixed points: the melting point of ice (or the freezing point of water) and the boiling point of water. Fahrenheit designated *32* as the freezing point of water and just as arbitrarily set *212* as its boiling point. The Fahrenheit scale thus divides the interval from 32 to 212 into 180 equal parts called degrees (denoted with the symbol °).

Few of us worry about the arbitrariness of this scale, especially if we are accustomed to it. Over the course of years we have learned to translate scale readings into "feel"; we automatically know the "feel" of various temperatures. We know, for example, that temperatures between 68°F and 72°F are comfortable indoor temperatures; 80°F is quite warm, but most welcome following a long winter; a temperature of 100°F is *hot;* a 50°F reading requires a light outer garment; and 98.6°F is the average normal body temperature.

The United States is one of the few countries that measures temperature in degrees Fahrenheit. In continental Europe and the rest of the world, temperature is measured in *degrees Celsius* or *centigrade (°C)*. The Celsius temperature scale was also named after its creator, the Swedish astronomer, Anders Celsius (1701–1744), who devised it in 1742 (about 28 years after Fahrenheit had devised his scale).

The freezing point of water was set at *0* on the Celsius scale and the boiling point of water was set at *100* so that there could be 100 equal subdivisions (called degrees Celsius) between them. The convenience of one hundred subdivisions made this scale ideal for scientific uses, and it has long been the standard scale of scientific temperature measurement. For a long time in English-speaking countries, the Celsius scale was called the centigrade scale ("cent" meaning 100 divisions or parts), but in 1948, the Ninth General Conference on Weights and Measures decided to abandon the name centigrade and use only "Celsius," in part to honor its creator.

If you've grown up using the Celsius scale, you probably think as little about its nature as those who have grown up with Fahrenheit think about their scale. Instead, you know what the temperature readings *feel* like: 20°C is quite comfortable, 37°C is normal body temperature and −5°C (which is read as "minus five degrees Celsius") would be the reading on a cold, but not unusually bitter winter day.

Here are some reference points tying the Fahrenheit and Celsius scales together:

$$98.6°F = 37°C$$
$$68°F = 20°C$$
$$23°F = -5°C$$

It's as difficult for a European to picture Fahrenheit degrees as it is for an American to get the feel of Celsius. But, when we convert to metrics, we'll all have to make the change to Celsius. Even now, more and more temperature readings in the United States are reported in degrees Celsius.

There are formulas that can be used to convert *exactly* any Fahrenheit temperature to a Celsius temperature. (There's also a formula for doing the exact conversion from Celsius to Fahrenheit.)

FAHRENHEIT → CELSIUS: *EXACT*

The exact conversion formula is:

$$C = \frac{5}{9}(F - 32)$$

In this formula:

C stands for degrees Celsius.
F stands for degrees Fahrenheit.
The parentheses tells us to complete this operation first, and the multiplication sign is implicit.

If we wanted to convert 75°F to Celsius, we would *first* substitute 75 for F in the formula, like this:

$$C = \frac{5}{9}(75 - 32)$$

$$= \frac{5}{9} \times \frac{43}{1}$$

75 minus 32 = 43. Next we made 43 a fraction by dividing it by 1 which does not change its value ($43 = \frac{43}{1}$).
Then we inserted the multiplication sign.

The final step involves carrying out the multiplication:

$$= \frac{5}{9} \times \frac{43}{1}$$

$$= \frac{215}{9}$$

$$= 23.9$$

So, 75°F is equivalent to 23.9°C, or 24°C rounded.

A Fahrenheit degree is *smaller* than a Celsius degree. It takes more of them, precisely 1.8 Fahrenheit degrees, to make one Celsius degree. This is because a Fahrenheit degree measures a smaller amount of *change*. There are 180 Fahrenheit degrees between freezing and boiling $(212 - 32 = 180)$, as compared with 100 Celsius degrees between the same two points $(100 - 0 = 100)$. This gives a ratio of 180 to 100, which is:

$$\frac{180}{100} = \frac{18}{10} = \frac{9}{5} = 1.8$$

But for *approximation* purposes, when the exact conversion is not important, we'll use 2 Fahrenheit degrees to make one Celsius degree (1.8 is close to 2.)

Here's how

FAHRENHEIT → CELSIUS: *APPROXIMATE*

If a Fahrenheit degree is smaller than a Celsius degree, it follows that a Celsius degree is "larger" (there are fewer of them to account for the distance between the freezing and boiling points of water.) One-half of a Celsius degree is equal to about one Fahrenheit degree. Actually, the exact ratio, the inverse of the ratio above, is:

$$\frac{100}{180} = \frac{5}{9} = 0.55$$

So, the *exact* relationship is 0.55 Celsius degree to each one Fahrenheit degree, but 0.50 or ½ is close enough for approximation. (Incidentally, we can substitute 0.55 for ⅝ in the exact conversion formula and do the computations and not involve fractions at all.)

Now, look back at the example we used to convert Fahrenheit to Celsius where 75°F was shown to equal 23.9°C. The result of 23.9°C makes sense since 68°F was already given to be equivalent to 20°C, and an increase of 7°F (from 68°F to 75°F) is equivalent to a rise of about 3.5 degrees Celsius (½ of 7). So, $20 + 3.5 = 23.5$, or about 24°C.

Let's do an example.

EXAMPLE: Find the approximate Celsius temperature corresponding to 50°F.

SOLUTION:

    (1) Remember that 50°F is 18° above freezing $(50 - 32 = 18)$.

    (2) Eighteen Fahrenheit degrees are roughly equivalent to about 9 Celsius degrees (½ of 18). So, in Celsius measurement, the temperature is approximately 9° above freezing.

(3) Freezing is 0°C. Therefore, 0°C + 9°C = 9°C, which is the approximate Celsius equivalent of 50°F.

The exact Celsius equivalent is $10° \left[\frac{5}{9} (50 - 32) \right]$ but, practically, the approximation is close enough.

✓✓*HOT TIP*   Another, fast way of obtaining a rough estimate is to *first* subtract 30 from the Fahrenheit reading, and *then* divide the answer in half. Like this:

50°F − 30 = 20°F
20°F ÷ 2 = 10°C

(We subtracted 30 rather than the 32 that appears in the exact conversion formula because 30 is an easier number to work with mentally. Dividing the answer in half is done because, as we explained, there are about 2 Fahrenheit degrees to every Celsius degree.)

EXAMPLE: Find the approximate Celsius temperature that is equivalent to 25°F.
SOLUTION:
 (1) Subtract 30° from 25° = −5°F.
 (2) Divide −5°F by 2 (divide in half) = −2.5°C or −3°C. (The exact Celsius equivalent is −3.89° or −4°C.)

Now, let's practice converting from Celsius to Fahrenheit, which is what we'd be doing in most travels outside the country.
Here's how

CELSIUS → FAHRENHEIT: *EXACT*

The exact conversion formula is:

$$F = \left(\frac{9}{5} \right) C + 32$$

$$F = 1.8C + 32$$

If you want to convert 15°C to Fahrenheit exactly, first substitute in the formula:

$$F = (1.8 \times 15) + 32$$

Then complete the calculations:

$$= 27 + 32$$
$$= 59$$

So, 15°C = 59°F.
    Now let's see what result we get by approximation.
    Here's how

CELSIUS → FAHRENHEIT: *APPROXIMATE*

In converting 15°C to Fahrenheit, recall that one of the reference points given was 20°C = 68°F. In approximating the conversion, we'll use the 20°C reference point because it's fairly close to the value (15°) we are considering.

Step 1 Note that 15°C is 5°C below 20°C.
Step 2 The 5 Celsius degrees are equivalent to about 10 Fahrenheit degrees. (There are about 2 Fahrenheit degrees to each Celsius degree, as you remember.)
Step 3 Since we are 5°C below 20°C, we should be about 10°F below 68°F. Thus, 20°C is about 58°F.

This certainly compares very well to the exact value of 59°F.
    ✓*TIP* Another fast way of estimating Fahrenheit degrees for a given Celsius reading is to double the Celsius figure and then add 30. (Again, we use 30 because it's somewhat easier to work with than 32; and we *double* the Celsius reading because each Celsius degree is equivalent to about two Fahrenheit degrees.)

EXAMPLE: Convert 34°C to Fahrenheit.
SOLUTION:
    (1) Double the Celsius reading ($34 \times 2 = 68$).
    (2) Add 30 ($68 + 30 = 98$).

Thus 34°C is about 98°F. Actually, the *exact* conversion is 93.2°F. This quick-and-dirty technique for estimating can produce a fairly substantial error for high temperatures. However, for most practical purposes, it's a very useful little trick. It certainly gives you the information you need to make most everyday decisions. After all, 93°F is as much of a beach day as 98°F would be!

Obviously, temperature conversion formulas are needed if you must have an exact temperature equivalent. However, aside perhaps from body temperature, this is generally not necessary in everyday life, and approximations will do just fine. To do approximations, it's handy to memorize some reference points, such as:

20°C = 68°F
 0°C = 32°F

You'll also need one or two facts:

One Celsius degree is about 2 Fahrenheit degrees.
One Fahrenheit degree is about ½ of a Celsius degree.

With this information firmly in mind, you can happily take off for distant lands knowing you'll need to take along a jacket if temperatures range from 15°C to 18°C.

Section 3: Let's Go Metric: Measurement Conversion

The U.S. system of weights and measures is an adaptation of the British system. For example, our inches, ounces, and pounds are directly inherited from the British, although our gallon is only about ⅚ of the British Imperial gallon.

Most other countries, especially those outside the British sphere of influence, use *metric measurements,* which is also the measurement system of choice for scientific purposes throughout the world. As more and more goods are manufactured for worldwide consumption, eventually everyone will be on metrics. Breaking with hundreds of years of tradition, Canada has recently converted to metrics and, sooner or later, and probably with great resistance, the United States will "go metric" as well.

We are used to the British (or American) system of measurement. We have the same "feeling" for the distance represented by a "mile" that we have for temperatures expressed in degrees Fahrenheit. Nevertheless, metrics is a more *rational* system. In metrics, all measures of length, weight, and liquid volume are built on the *decimal* system. *All quantities are expressed in 10's or multiples of 10.* This makes computations and comparisons relatively easy.

In this section we will show you how to convert to and from metrics and, more important, also give you a "feel" for metric units in much the same way that we did with "degrees Celsius."

Here are some things you need to know. In metrics:

Milli means one-thousandth ($\frac{1}{1000}$)
Centi means one-hundredth ($\frac{1}{100}$)
Deci means one-tenth ($\frac{1}{10}$)
Kilo means one thousand (1,000)
One kilometer = $\frac{5}{8}$ (.625) mile (approximately)
One meter = 39.37 inches
One inch = 2.54 centimeters
One mile = $\frac{8}{5}$ (1.6) kilometers (approximately)

The language of the metric system provides hints for converting *within* the metric system. For example, since "centi" means one-hundredth ($\frac{1}{100}$), one centimeter equals $\frac{1}{100}$ meter, and 250 centimeters equals $\frac{250}{100}$ meters (which is equivalent to 2.5).

As another illustration, consider milligrams. One milligram equals $\frac{1}{1000}$ gram, so 325 milligrams equals 0.325 grams. (You arrive at this by dividing 325 by 1000.)

The conversion of kilograms to grams also flows from the language of metrics. Since one kilogram equals one thousand (1,000) grams, 2.5 kilograms (2.5 thousand grams) equals 2,500 grams ($2.5 \times 1,000$).

Now, let's convert grams to kilograms. Because one kilogram equals 1,000 grams, 350 grams is equal to $\frac{350}{1000}$ kilograms or 0.35 kilograms.

The examples we just did focus on conversions within the metric system. In the rest of this section, we will convert between the metric and British system.

Here's how

MEASURING LENGTH

"How tall are you, Antoine?"

"Oh, I'm about 180 centimeters."

"That's what I thought—about average. It's your brother, Pierre, who's really tall . . . I'd guess about 2 meters."

This conversion makes perfect sense to a Parisian, although it may not to you. Yet, if you had to guess at Antoine's height from the little hints, you might say that his "about average" height of 180 centimeters is about 5 feet 10 inches, the average height of a man in the United States. Similarly, you'd probably consider 6 feet 6 inches "really tall."

Converting from centimeters and meters to inches and feet (and vice versa) requires a few definitions and facts. The actual conversion process takes only a few moments.

To Convert Centimeters or Meters to Inches *Exactly:*

Step 1 (If you're starting with meters, skip Step 1 and just go right on with Step 2.) Since centi- means one-hundredth, first convert the number of centimeters to meters by dividing by 100 (move the decimal point 2 places to the left).

Step 2 Since one meter = 39.37 inches, convert the meters to inches by multiplying the number of meters by 39.37.

Let's do an example based on Antoine's height. We'll convert 180 centimeters to feet and inches, like this:

180 cm (centimeter) = 1.80 m (meter) $\begin{cases}$ Divide by 100, or move the decimal point 2 places to the left. $\end{cases}$

= (1.80 × 39.37 in) (inches)
= 70.866 in

Now dividing by 12 inches (since there are 12 inches in a foot), you find that 70.866 is about 5.91 feet. Next we note that 0.91 feet represents about 11 inches (0.91 × 12 inches). Thus, 70.866 inches is about 5 feet 11 inches. (Recall that based on guesswork alone we estimated that 180 cm was equivalent to about 5 feet 10 inches.)

As a second example, convert 2 meters (m) to feet and inches.

2m = (2 × 39.37 in)
= 78.74 in

That's about 79 inches, or 6 feet 7 inches—pretty close to our intuitive estimate for Pierre of 6 feet 6 inches.

EXAMPLE: While watching the Olympic track and field events, you became curious about exactly how far 400 meters is in yards.
SOLUTION:
Step 1 Convert the meters to inches by multiplying 400 by 39.37 because there are 39.37 inches per meter:
(400 × 39.37 in) = 15,748 in
Step 2 Then divide by 36 because there are 36 inches in a yard:
(15,748 ÷ 36 yards) = 437.4 yards

By calculator:
PRESS 400 ⊠ 39.37 ÷ 36 ⊟

The 400 meter run is thus quite close in length to the 440 yard run which is, in fact, one-quarter mile. (There are 5,280 feet per mile, which when divided by 3 feet per yard equals 1,760 yards. Dividing 1,760 yards by 4 (quarters of a mile) equals 440 yards.)

✓*TIP*   For *approximation purposes,* think of the meter as an over-sized yard since 1 meter = 39.37 inches = 3 feet 3 inches (approximately), or about 1 yard and 3 inches. So, if a sign announces an exit in 100 m, it's reasonable to think of that distance as a bit longer than 100 yards.

To Convert Kilometers to Miles:
Since 1 kilometer equals ⅝ (0.625) mile, multiply the number of kilo-meters by 0.625 to get the equivalent number of miles.

EXAMPLES:

5 km (kilometer) = (5 × 0.625) miles = 3.125 miles
15 km = (15 × 0.625) miles = 9.375 miles

✓*TIP*   If you want to do an easier approximate conversion from kilo-meters to miles, remember that 1 kilometer is close to two-thirds of a mile (⅔ = 0.667). Therefore, multiply the number of kilometers by ⅔. An easy way to do this is to *divide the number of kilometers by 3 and multiply the result by 2* to obtain the number of miles:

15 km = [(15 ÷ 3) × 2] miles = 10 miles

The exact equivalent we just saw was 9.375 miles.

Or, equivalently, *multiply the number of kilometers by 2 and divide the result by 3* to obtain the approximate number of miles:

15 km = [(15 × 2) ÷ 3] miles = 10 miles

These conversions were *from* the metric system *to* the Imperial (or U.S.) system. Converting in the other direction, *to* metrics, is no more (or less) complicated.

To Convert Inches, Feet or Yards to Centimeters or Meters:

Step 1   First, convert the number of feet to inches (by multiplying by 12) or convert the number of yards to inches (by multiplying by 36).
Step 2   Since 1 inch = 2.54 cm, multiply the number of inches by 2.54 to get the number of centimeters.

Step 3 If desired, the number of centimeters can be converted to meters by moving the decimal point 2 places to the left (which is the same as dividing by 100).

EXAMPLE: Convert 5 feet to centimeters, then to meters.

$$5 \text{ ft} = (5 \times 12 \text{ in}) = 60 \text{ in}$$
$$= (60 \times 2.54) \text{ cm} = 152.4 \text{ cm}$$
$$= 1.52 \text{ m}$$

✓*TIP* To do an *approximate* conversion from inches into centimeters: *Divide the number of inches by 2 and multiply the result by 5, or multiply the number of inches by 5 and then divide the result by 2.*

$$25 \text{ in} = [(25 \div 2) \times 5]\text{cm} = 62.5 \text{ cm},$$
$$\text{or}$$
$$25 \text{ in} = [(25 \times 5) \div 2]\text{cm} = 62.5 \text{ cm}$$

The *exact* conversion of 25 inches to centimeters $= (25 \times 2.54)$ cm $= 63.5$ cm). The approximation technique works because 2.54 is about equal to 2.5 and 2.5 equals 5 divided by 2. Multiplication by 2.54 is approximately the same as multiplication by $5/2$ which is done by first multiplying by 5 and then dividing by 2.

To Convert Miles to Kilometers:
Multiply the number of miles by 1.6:

$$60 \text{ miles} = 60 \times 1.6 \text{ km} = 96 \text{ km}$$

✓*TIP* A rough estimate of the number of kilometers can be obtained by multiplying the number of miles by 3 and dividing the result by 2 (or by dividing the number of miles by 2 and multiplying the result by 3). This works because 1.6 is about $1.5 = 3/2$.

$$60 \text{ miles} = [(60 \times 3) \div 2] \text{ km} = 90 \text{ km}$$
$$\text{or}$$
$$60 \text{ miles} = [(60 \div 2) \times 3] \text{ km} = 90 \text{ km}$$

We employ similar procedures when we want to convert weights to and from the metric system.

Here's how

MEASURING WEIGHTS

As with lengths, there is both a British (or American) system of weights and a metric system. While we and the British are familiar with pounds and ounces, in Europe, one would commonly buy ½ kilo of tomatoes, 100 grams of cheese, or 5 kilos of potatoes. The typical woman weighs 50 to 60 kilograms, while the average man's weight is between 60 and 85 kilograms. Even in the United States, the standard dosage of an aspirin tablet is quoted as 325 milligrams. These quantities lose their mystery when you know some additional metric facts:

> *Kilo* means one thousand
> *Milli* means one-thousandth
> *One kilogram* = 2.2 pounds
> *One pound* = 454 grams

To Convert Kilograms to Pounds and Ounces:

There are two steps for doing an *exact* conversion from kilograms to pounds and ounces:

Step 1 Since 1 kilogram (kg) = 2.2 pounds (lbs), multiply the number of kilograms by 2.2 to obtain the number of pounds as a decimal. Note that the decimal portion of your answer is *not* ounces, but a fraction of a pound. For example,

56 kg = (56 × 2.2 lbs) = 123.2 lbs.

Step 2 Multiply the decimal portion of the number of pounds by 16 to obtain the number of ounces. (Remember, there are 16 ounces in a pound.) Continuing our example, 0.2 × 16 ounces = 3.2 ounces. So, 56 kilograms is about 123 pounds 3 ounces.

EXAMPLE: Convert ½ kilogram to pounds.

½ kilo = 0.5 × 2.2 lbs = 1.1 lb (Notice that one pound is about the same as ½ kilo! That's why it's reasonable to buy ½ kilo tomatoes.)

✓*TIP*   You can think of 1 pound and ½ kilo as roughly equivalent for most practical purposes.

EXAMPLE: Convert 100 g (g = grams) to ounces.

100 g = ¹⁄₁₀ kg, since 1 kilogram = 1,000 grams

100 g = ¹⁄₁₀ kg = ¹⁄₁₀ × 2.2 lbs = 0.22 lbs. (This answer is in *hundredths* of a pound. Convert it to ounces by multiplying by 16.)

0.22 lbs = 0.22 × 16 oz (ounces) = 3.52 oz

Thus, 100 grams of cheese is close to one-quarter pound of cheese (4 ounces).

✓*TIP*   A handy rule of thumb is that 100 grams approximates one-quarter of a pound.

EXAMPLE: Convert 325 mg (milligrams) to ounces.

First, since "milli" means one-thousandth ($\frac{1}{1000}$), to convert to grams, divide by 1,000:

$$325 \text{ mg} = {}^{325}/_{1000} \text{ g} = 0.325 \text{ g}$$

But one kilogram is 1,000 grams, so one gram is $\frac{1}{1000}$ kilogram.

Then $0.325 \text{ g} = {}^{0.325}/_{1000} \text{ kg} = .000325 \text{ kg}$

This is a very small quantity, but, continuing our conversion, we have:

$$0.000325 \text{ kg} = 0.000325 \times 2.2 \text{ lb} = 0.000715 \text{ lb}$$
$$= 0.000715 \times 16 \text{ oz} = .01144 \text{ oz}$$

Thus, a typical aspirin tablet weighs about one-hundredth (0.01) of an ounce, a *very* small quantity indeed! But it is *not* necessarily a small amount of medication. Medication is generally given in very small doses, usually measured in milligrams rather than grams or ounces.

✓✓*HOT TIP*   You can convert kilograms to pounds *approximately* by multiplying the number of kilograms by 2 (rather than by 2.2).

To Convert Pounds to Kilograms:

For an *exact* result, divide the number of kilograms by 2.2, since there are 2.2 pounds per kilogram:

$$155 \text{ kg} = (155 \div 2.2) \text{ lbs} = 70.5 \text{ lbs}.$$

✓*TIP*   To *estimate* how many pounds there are in a given number of kilograms,   number of kilograms by 2. The approximate conversion would be $155 \div 2 = 77.5$ pounds. The result is higher than the exact answer because we are dividing by a lower number (2 rather than 2.2) than we should be.

To Convert Ounces to Grams:

This conversion requires two steps for an *exact* answer:

Step 1 First convert ounces to pounds by dividing the number of ounces by 16.

Step 2 Multiply the Step 1 result by 454 since there are 454 grams in a pound.

EXAMPLE: Convert 8 ounces to grams.

$8 \text{ oz} = {}^{8}/_{16} \text{ lb} = 0.5 \text{ lb}$ (Note that 8 ounces = one-half pound.)

$= 0.5 \times 454 \text{ g} = 227 \text{ g}$

EXAMPLE: Convert 5 ounces to grams.

$5 \text{ oz} = \frac{5}{16} \text{ lb} = 0.3125 \text{ lb}$
$= 0.3125 \times 454 \text{ g} = 142 \text{ g}$

Of all the metric measures, none has had as great an impact on American society as the measurement of liquid volume. Pepsi, Coke, and other soft drinks now come in one-liter or two-liter bottles, and cans are marked in both ounces and milliliters.

✓*TIP*   If you aren't already accustomed to doing so, think of a liter as being a bit more than a quart and two-liter bottles as containing somewhat more than a half-gallon (two quarts).

Here's how

MEASURING LIQUID VOLUME

In the metric system the liter is the basic unit of volume. By definition,

1 liter = 1,000 cubic centimeters (1,000 cc).

In the American system, the *exact* relation between liters, quarts, and ounces is:

1 liter = 1.0567 quarts
1 liter = 33.8 ounces
1 quart = 0.946 liter
1 ounce = 0.03 liters or 33 cc

You can see that a liter is really close to a quart (qt).

To convert liters to quarts, you have to keep in mind that the quart is divided into liquid ounces. There are 32 ounces in one quart.

To Convert Liters to Quarts and Ounces:

There are two steps in an *exact* conversion:

Step 1 Since 1 liter = 1.0567 quarts, multiply the number of liters by 1.0567 to get the exact number of quarts.

Step 2 To convert the decimal part of the number of quarts to ounces, multiply the decimal by 32, since there are 32 ounces in one quart.

EXAMPLE: Convert 2 liters to quarts.
(1) 2 liters × 1.0567 qts per liter = 2.1134 qts
(2) 0.1134 qt × 32 oz per qt = 3.63 oz
2 liters = 2 qts and 3.63 or about 4 oz

To convert liters to quarts *approximately:*

Step 1 Multiply the number of liters by 1.1 (instead of 1.0567).
Step 2 Multiply the decimal part of the number of quarts by 32.

EXAMPLE: Convert 2 liters to an approximate number of quarts and ounces.
(1) 2 liters × 1.1 qt per liter = 2.2 qts
(2) 0.2 qts × 32 oz per qt = 6 oz (approximately)
2 liters = 2 qts and about 6 oz

This is a little high in comparison with the exact answer of 2 quarts 3.6 ounces, but it is good enough for most practical purposes.

Dosages of liquid medicines are often given in cc (cubic centimeters) or, equivalently, in ml (milliliters). Notice that, since

1 liter = 1,000 cc,
1 cc = $1/1000$ liter = 1 ml

To Convert Cubic Centimeters or Milliliters to Ounces Exactly:
Note that since 1 liter = 33.8 oz, 1 cc = 33.8 oz/1,000 = 0.0338 oz. To find the number of ounces:
Multiply the number of cc's by 0.0338 to obtain the exact number of ounces.

EXAMPLE: The doctor prescribes 10 cc of medicine every 3 hours. How much is this in ounces?
10 cc × 0.0338 oz per cc = 0.338 oz, or about $1/3$ oz
10 cc are about $1/3$ oz.

You might also be interested in these facts:

One tablespoon = $1/2$ oz
One teaspoon = $1/3$ tablespoon = $1/6$ oz

Because 2 teaspoons are $2/6$ (or $1/3$) oz, 10 cubic centimeters equal about 2 teaspoons!

To Convert from Quarts and Ounces to Liters Exactly:

Step 1 Since 1 quart = 0.946 liters, multiply the number of quarts by 0.946.

Step 2 Since 1 oz = $\frac{1}{32}$ qt = 0.03125 qt, first multiply the number of ounces by 0.03125 to get the number of quarts. Then, to convert these quarts to liters, multiply by 0.946 (because 1 qt = 0.946 liters).

EXAMPLE: Convert 2 quarts 3 ounces to liters.
3 oz = 3 × 0.03125 qts = 0.0938 qts
Altogether, 2.0938 qt × 0.946 liter per qt = 1.98 liter.

That's about all you need to know to convert between the American and metric measurements. With these simple facts and conversion methods at your fingertips, you need never again by mystified by the metric system.

Section 4: Electrical Conversion, or Will My Hair Drier Blow Abroad?

You're about to start on your first vacation abroad. In preparation, you've bought some new clothes, refilled all your cosmetics and prescriptions, bought a travel iron, and made a note to pack your electric razor and blow drier.

In truth, it's not necessary to pack much differently for a vacation abroad than for an American one. Soap, cosmetics, and all kinds of nice clothes are available from Sweden to Singapore, so if you forget something or run out of it, buying a replacement is really no problem. And shopping abroad for odds and ends like toothpaste, anti-acid tablets, or band-aids can be quite an amusing adventure. Nevertheless, you'll probably want to pack everything you think you might need.

If you read the beginning chapters of your guidebook, you'll have learned that in most foreign countries, the standard voltage for electrical current is 220–240 volts. In the United States, it's 110–120 volts. This difference in voltage means that, all else being equal (and it's not—the plugs and sockets are different too), electrical equipment designed to run at 110–120 volts will *not* work at 220–240 volts. The motor will burn out. Similarly, 220–240 volt equipment can't be used on 110–120 volt lines.

So what do you do about the iron, razor, and hair dryer? You actually have a few different options.

Here's how

CONVERTING VOLTAGE

Your object is to use the correct appliance on the appropriate electrical system. First, find out the voltage requirement of the appliance. All electrical appliances have this information printed on them, usually on the bottom or back of the device near the power cord. Electrical appliances can usually tolerate a 10% deviation from the quoted voltage requirements and still function effectively. Thus, if your TV requires 120 volts, it will work just fine with only 110–115 volts. It will not work, however, in Europe where the 220–240 volt system applies. (Note that not all foreign countries are on 220–240 volts. Check a guidebook for the particular country you're interested in.)

The 220–240 volt system allows you to obtain the same wattage (electrical power) with half the current required in the 110–120 volt system. But, on the other hand, the 220–240 volt system is a bit more dangerous in that you'd receive a more severe electrical shock should you accidentally put your finger in a live socket. If you would like to learn a little more about electricity, skip ahead to Chapter 12, Section 2.

One way of making sure that your appliance can be used in a foreign country is to buy a *converter*. This is a fist-sized gadget that converts 220–240 volt current to 110–120 volts. You can use the same converter for all the electrical equipment you have.

Another alternative is to buy new appliances especially designed for foreign travel. This sort of equipment has special circuits built in that allows you to run on either voltage system at the flip of a switch. As more and more people do more and more traveling, these dual-system appliances are becoming increasingly common, although you may still have some difficulty finding one. Your local appliance store may be able to order these appliances for you from the manufacturer or wholesaler.

Alternatively, you can purchase electrical equipment made for the foreign market that works on 220–240 volts exclusively. That's what we did in Paris last year when our hair drier broke—but, remember, these appliances can't be used on 110–120 volt systems back home.

✓*TIP* Other than in large U.S. cities, 220–240 volt appliances will be very difficult to find; so, if you really want a piece of equipment just to use in a specific country, buy it there. They may also have plug configurations that can't be used on 220–240 volt sockets.

If you travel extensively, it probably makes good sense to buy 220–240 volt appliances or dual-purpose ones. Consider these options carefully, especially when you're thinking of replacing your regular equipment with lighter-weight, travel-sized models.

So, packed and armed with your electric voltage converter you make the trip to Rome. After arriving, quite tired, and settling down in your hotel, you decide to take a shower and wash your hair. A refreshing shower perks you up and you get out your hair drier . . . to find you can't plug it into the wall!

Here's how

> ## ADAPTER PLUGS

Not only is the voltage system different in Europe, the outlets (sockets) are different too, a fact that many travel guidebooks fail to warn you about. Your made-in-America appliance, whether the 110–120 volt one with a converter or the dual-system one with its built-in converter, ends in a plug that looks like this:

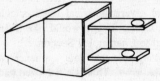

and all the sockets at home look like this: or this:

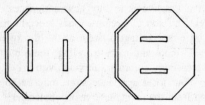

except for the 3-pronged heavy duty ones:

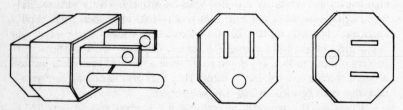

To use your made-in-America electrical appliance abroad, you need a *plug adapter* in addition to a converter. This is a gadget that looks like a plug without a cord. On the end where you'd expect to find the power cord, you'll find a socket into which you fit the plug of your appliance. The other end goes into the wall outlet.

There are a variety of plug ends that go into the various international wall outlets—and there are different ones for different countries. They may look like these:

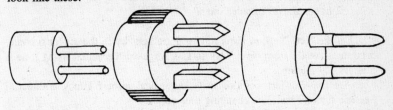

with different length prongs and spacing between the pins.

Wall outlets in most countries abroad look like this and require the type of plug that has thin, round pins:

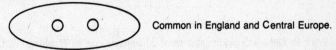

Common in England and Central Europe.

These are some of the other common outlet configurations; they correspond to the plugs illustrated above:

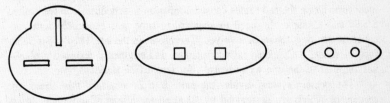

England/Africa/Hong Kong   Caribbean/South America   Europe/Africa/Asia/Middle East

For a few dollars you can buy a 4- or 5-piece plug set useful worldwide. Remember, though, that adapter plugs do not affect the voltage. They only enable you to connect your appliance into the electrical system. You'll probably need both an adapter plug *and* a voltage converter to make your blow drier blow—and not blow out its motor.

Section 5: Changing Time Zones

When traveling in the United States or abroad, you invariably have to deal with plane or train or bus schedules. Here are a few typical examples of the kind of situations you're likely to run into:

- You leave New York at 7:30 PM, fly by commercial jet at a cruising speed of about 600 miles per hour, and arrive in Paris (3,000 miles east) at 7:35 AM—it looks like 12 hours and 5 minutes later.
- A plane leaves New York at 11:00 AM, travels west 3,000 miles to California, and sets down in Los Angeles at 1:29 PM—is it really only 2 hours and 29 minutes later?

On the New York to Paris run, you are actually in the air for 6 hours, 5 minutes, while from New York to Los Angeles, the actual flying time is 5 hours, 29 minutes.

Are these differences due to typographical errors? Fancy arithmetic? Science fiction? No . . . changing time zones.

Here's how

STANDARD TIME

Prior to 1884 when delegates from 27 nations met in Washington, D.C. and agreed to adopt a worldwide uniform time system, each local community kept its own time. The need for a standard system arose with the development of rapid railway transportation in the 19th century (differences in local time along the rail routes caused confusion in schedules). In the United States and Canada, the need to standardize time was especially acute because of the long distances involved. Therefore, in the late 1870's, Sir Sandford Fleming, a Canadian railway planner and engineer, designed a plan— very similar to the one we now use—for worldwide standard time.

The present system divides the earth into *24 standard time zones,* the centers of which are designated by standard meridians of longitude, imaginary lines running from the North Pole to the South Pole that meet the equator at right angles. The equator, which is also an imaginary line around the earth, measures 24,000 miles in circumference and is equidistant at all points from the North and South Poles.

These 24 meridians of longitude are set 15 (spherical) degrees (15°) apart at the North Pole, starting with the prime meridian (of 0°) at Greenwich, England. You can visualize this by thinking of how the sections of an orange meet at the stem. Any section designated as the starting place would be equivalent to 0°, like Greenwich, England. As you may remember from geometry, there are 360° in a circle. The 24 meridians of longitude multiplied by 15 degrees for each meridian equals 360 degrees, the number of degrees in an imaginary circle drawn at the top of the earth with the North Pole as center. (See the drawing on the next page.)

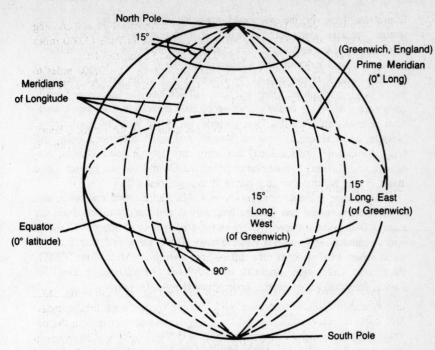

The 24 meridians of longitude (denoted by the broken lines from the North to the South Poles) do not form the boundary lines of the standard time zones, but rather, the zones' *theoretical centers*—like this:

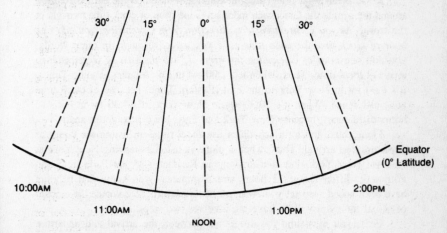

In practice, however, the east-west boundaries of the time zones do not run straight but have been altered in shape, or subdivided in some cases, for the convenience of the inhabitants. For example, the time zone that would bisect Siberia has been redrawn with a bulge to take in all of Siberia.

If you look at the map of the United States in the front of the white pages of your telephone directory, you will see how the time zone boundaries zigzag down the country—not necessarily coincident with state boundaries. For example, Pacific Time includes all of the states of Washington, Nevada, and California, most of Oregon, and small corners of Idaho and Utah. Mountain Time's easterly boundary cuts through South Dakota, Nebraska, and Kansas. Similar warps define the Central and the Eastern Time zones. These are the four time zones of the continental U.S.

Clock time is the same everywhere within a zone (north and south, and to the zone's eastern and western boundaries) and usually differs from the starting base point at Greenwich by an *integral number* of hours (i.e., *whole hours*—minutes and seconds don't count). There are a few places in the world where local standard time differs from Greenwich Mean Time (GMT), the time at Greenwich, England, by half-hours (Afghanistan, Burma, Sri Lanka, for example) or quarter-hours (for example, Guyana).

Here's how

TRAVELING EAST

Local mean solar time (time measurement based on the earth's rotation around the sun) at the Greenwich meridian is the basis of clock time throughout the world. *As you go in an easterly direction from Greenwich, you add one hour of clock time for each time zone you travel through.* In other words, you add one hour *of clock time for every 15° to the east of Greenwich* to arrive at local time. Thus, when it's 1:00 PM in Los Angeles (Pacific Time), it's 4:00 PM in New York to the east (Eastern Time) because of the 3 time zone difference. When it's 10:30 AM in New York, it's 4:30 PM in Paris (6 hours difference), because New York and Paris are 6 time zones apart.

Plane, bus, and train schedules use *local* times in listing the hours of departure *and* arrival. Thus, when a plane is listed as leaving New York at 7:30 PM New York time and arriving in Paris at 7:35 AM Paris time—an arithmetic difference of 12 hours and 5 minutes—you know that 6 hours have been added (you set your watch 6 hours *ahead*). To compute the amount of actual time you will be traveling involves two steps:

First, you must find the difference between the arrival and departure times (which means that you have to know how to subtract and add hours and minutes); then you must adjust for the time zone changes.

As an example, let's use the New York to Paris flight where you leave New York at 7:30 PM and arrive at Paris at 7:35 AM. How long is this interval?

Subtract 7:30 PM from 7:35 AM, like this:

From 7:30 PM until midnight   = 4 hrs 30 min
+   +
From midnight to 7:35 AM   = 7 hrs 35 min
                    Total clock time = 11 hrs 65 min = 12 hrs 5 min

The total difference is 12:05. Then, because you were going in an easterly direction through 6 time zones (for which you *added* 6 hours), you *deduct* these 6 hours for the *actual* time you'll be airborne:

    12:05 (local standard time difference)
  −6:00 (time zone difference)
    6:05 of actual traveling time

When going in a westerly direction, you compute time in much the same way.

Here's how

TRAVELING WEST

When you go in a westerly direction from Greenwich, to compute local time, you subtract one hour for every 15° of longitude.

So, because New York and California are 3 time zones apart, there is a 3 hour time difference: if it's noon in New York, you subtract 3 hours to compute California time at 9:00 AM. When it's 11:00 AM in New York, it's 6:00 AM in Hawaii, because Hawaii is 5 time zones away in a *westerly direction*. Therefore "time" is set *back* that many hours.

Now, let's figure out how many hours of actual flying time elapses during a plane trip going west from New York to Hawaii, via California:

From the earlier example, we know that the plane departs New York at 11:00 AM and arrives in Los Angeles at 1:29 PM.

Therefore:

From 11:00 AM to 12:00 noon = 1 hour
+   +
From 12:00 noon to 1:29 PM = 1 hr 29 min
                Total clock time = 2 hrs 29 min

However, you have to account for the 3-hour time difference. Since Los Angeles is 3 time zones to the *west* of New York, it is 3 hours "earlier" there. To compute the actual flying time, you have to *add* those hours back in, like this:

> 2:29 (standard time difference)
> + 3:00 (time zone difference)
> = 5:29 (actual flying time)

Suppose the plane leaves California at 9:00 PM and arrives in Hawaii at 1:29 PM

> 11:29 PM (arrival, Hawaii time)
> −9:00 PM (departure, L.A. time)
>   2:29 (local standard time difference)

Then, add 2 hours for the 2 time zones' difference between California and Hawaii to get:

> 4:29 minutes of actual flying time.

In total, it took 5:29 minutes to fly from New York to California and another 4:29 minutes to go from California to Hawaii. Total actual flying time is:

> 5:29
> +4:29
> 9:58

This does not count, of course, any layover wait between planes or other stops along the way.

If you circled the globe completely in either direction, east or west, you would travel through 24 time zones and, theoretically, gain or lose a whole calendar day. But this doesn't happen.

Here's how

INTERNATIONAL DATE LINE

The international date line is an imaginary line in the middle of the Pacific Ocean that follows the meridian of 180°—exactly halfway around the world from Greenwich, England—but adjusted so as to include the Aleutian Islands with Alaska and some of the South Sea Islands with Australia. Selected by mariners because it was "convenient," it is arbitrarily defined as the place where each new date begins. When traveling across the international date line going *west*, the date becomes a day later. Going *east* across the international date line, the date becomes a day earlier. To understand why a change of full day must occur somewhere, consider the following example.

Suppose you could travel around the earth in 2 minutes. If there was no international date line, you could leave Greenwich, England at noon on December 24, go east through 24 time zones and arrive back in Greenwich at 12:02 PM. This would be 12:02 PM on December 25, because when you arrived at the time zone in which local time was midnight, you would start the next day.

But this doesn't happen because a correction is made at the international date line. Here, halfway around the world from Greenwich (and traveling east), the calendar date is set back by one day (to December 24). Since we left at noon and it takes 2 minutes to travel around the earth, halfway around would be 12:01 AM. If there was no international date line, it would be 12:01 on the 25th. Instead, the international date line corrects this by setting the calendar back one day: it's still 12:01 AM, but it's 12:01 AM on December 24. So one more minute and 12 time zones later, you arrive safely back in Greenwich at 12:02 PM on December 24.

In flying from California to Australia (going west through 6 time zones), you cross the international date line. The time between a scheduled takeoff of 9:00 PM on a Tuesday and a scheduled landing of 7:35 AM on Thursday is computed as before:

From 9:00 PM to midnight = 3 hrs
+   +
From midnight to 7:35 AM = 7 hrs 35 min
                Total clock time = 10 hrs 35 min

Now add 6 hours (one for each time zone). The trip actually takes 16 hours and 35 minutes. The calendar date can be ignored since the clock times are independent of the date.

Moving clocks back and ahead are arbitrary decisions that have little to do with the earth's rotation on its axis (which defines a day and night and seasons) or with the earth's rotation around the sun. Daylight Savings Time is a good example.

Here's how

DAYLIGHT SAVINGS TIME

First proposed by Benjamin Franklin in 1784, used extensively during World Wars I and II, and finally adopted by an act of Congress in 1966 (the Uniform Time Act), daylight savings time is a way of using more daylight, especially in the summer months when more daylight is available. By setting the clock ahead in the summer, we have the advantage of more light later in the day.

In both the world wars, daylight savings time was in force as a means of saving fuel by reducing the need for artificial light. For a period of 3 years and 8 months during World War II, the United States stayed on daylight savings time year round.

Most states (not all—some use a variation or stay on standard time all year) follow daylight savings time and set the clock one hour ahead (of local standard time), usually some time in April. The clock is set back again one hour (to local standard time), usually some time in October.

✓*TIP*   There is a handy mnemonic device that tells you the direction to set the clock: "Spring *ahead*, Fall *back*."

Since plane, train, and bus schedules are always stated in local times, they will reflect daylight savings time adjustments for those places in the United States or abroad that use it. But, since not all countries are on daylight savings time, a place 5 or 6 time zones away that is not on daylight savings time may actually differ from your local time by only 4 or 5 clock hours in the summer months.

7

Sports
and Car Math

Section 1: Miles Per Gallon

When you buy a new car, information about expected gas consumption (mpg—miles per gallon) is included on the ticket. Yet surprisingly few people bother to compute the number of miles per gallon their car is actually getting on a regular basis, despite the fact that performing this easy computation can often enable you to detect early signs of engine trouble.

Several years ago, a friend of ours was driving his Volkswagen from San Francisco to New York when, just outside of Sioux Falls, South Dakota, he recorded a sudden drop in gas mileage. Thinking it might have been a fluke, he waited until the next time he filled up and again computed his miles per gallon. The figure was even lower than before. Nothing else seemed wrong, but he decided to stop at a VW dealer and have the car checked out—just in case. It turned out that one of the valves was badly burned. The dealer was most helpful. In fact, the mechanic came in at 4:30 the following morning to complete the repairs. If Jeff hadn't been alert to this early warning indicator, the VW would surely have suffered major engine damage.

Computing miles per gallon is relatively straightforward.
Here's how

MPG

Keeping track of your gas mileage involves noting the *odometer reading* each time you fill up your gas tank. Start now if you don't already do this habitually. And follow these easy steps:

Step 1 The next time you stop for gas, be sure to fill the tank up completely and make note of the odometer reading. To compute mpg's, each odometer reading *must* be taken when the tank is filled up.

Step 2 Drive as you normally would and stop for gas whenever you want to. But whenever you fill up, be *sure* to completely fill up the tank again.

Note the current odometer reading and the number of gallons needed to fill the gas tank.

Step 3 Figure out the difference between the current and the previous odometer readings. This number is the number of miles you have driven between successive fill-ups.

Since you started out with a full tank, the number of gallons needed to fill the tank again is the number of gallons you used in driving the number of miles since the last fill-up.

Step 4 Divide the number of miles between fill-ups by the number of gallons of gas used between fill-ups. The result is the *number of miles per gallon of gas (mpg)*.

These steps can be summarized with a formula:

$$\text{Mpg} = \frac{\text{New odometer reading} - \text{Previous odometer reading}}{\text{Number of gallons of gas needed to fill tank}}$$

EXAMPLE: Speedy filled up her tank in Boston when the odometer read 21,565. When she filled up the tank again in Philadelphia, she needed 13.4 gallons of gas. The odometer read 21,884. Find the number of miles per gallon of gas Speedy was getting on her trip south.

SOLUTION: Substituting in the formula, we have:

$$\text{Mpg} = \frac{21,884 - 21,565}{13.4}$$
$$= \frac{319}{13.4}$$
$$= 23.8$$

Speedy got 23.8 mpg's on the Boston to Philadelphia leg of her trip.

✓*TIP* To keep track of your mpg's, it's a good idea to keep a small notebook in your car. You can use it to record the date, odometer reading, number of gallons of gas to fill the tank, and the amount you paid for the gas—as well as for computing your gas mileage. Remember, you can only compute miles per gallon from fill-up to fill-up. For your car's sake, it's worth your while to make this effort.

Section 2: RPM's and Gear Ratios

Here's some information about:

RPM's

In the olden days, about 35 years ago, phonograph records were quite brittle. Needless to say, lots of records broke. And since they played at 78 rpm's, that is, they completed 78 *revolutions per minute* on the turntable, they revolved so fast the needle spiraled from the record's outer edge to the center in just a very few minutes. As a result, not too much music could be recorded on one record.

More recently, two types of records were produced: 45 rpm's and 33⅓ rpm's. The 45 rpm is smaller and has a big hole in the middle; it is most always used for "singles." Albums are recorded on the large 33⅓ rpm discs. Improved technology, including the use of superior materials, permitted the recording of as much as 20 to 30 minutes of music on each side of these slowly revolving discs. Today, of course, most music is being produced on cassette and compact disk.

Rotational speed is generally measured in revolutions per minute (or per second), or in degrees per minute (or per second). For example, since a full rotation around a circle is 360 degrees, a rotational speed of 360 degrees per minute would be equivalent to a rotational speed of one revolution per minute.

Many of today's cars are equipped with *tachometers,* which indicate how fast the drive shaft is being turned by the engine. For example, a reading of 6,000 rpm means the drive shaft is spinning at 6,000 revolutions per minute. Most tachometers have a red zone warning drivers of the maximum permissible rpm's before engine damage begins to occur.

Here's how gears work

GEARS

An automobile engine is connected to the car's rear wheels (or to the front wheels in a front wheel-drive automobile) by means of a drive shaft and gears. These effect the transfer of power from the turning of the engine to the turning of the wheels.

Gears have teeth that mesh. The ratio of the number of teeth on one gear to the number of teeth on the gear with which it meshes is called the *gear ratio*. On a bicycle, pedaling turns a front gear, which, in turn, is attached by the chain to a gear attached to the rear wheel. Pedaling produces *torque* (rotary force) in the front gear which is transmitted to the rear wheel.

When a small gear turns a larger gear, such as the gear on the back wheel of a bicycle, all of the rotary force created by pedaling one revolution of the small gear is transferred to the rear gear in such a way that the wheel moves only a little bit. Your best effort in this case only creates a little movement—the bike goes slowly. Carefully examine the gears of a 10-speed bicycle. You will discover that the "lowest" gear is when the gear in use at the pedal is smallest and the gear in use at the rear is largest: this is the most powerful, but slowest gear. Similarly, the highest gear is when the largest gear at the pedal turns the smallest gear at the rear.

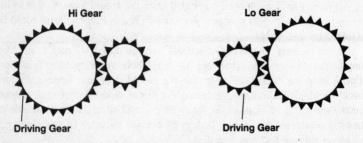

Hi Gear

Driving Gear

Lo Gear

Driving Gear

The same principle also applies to automobiles. In first gear, which is the lowest gear, a small gear (the driving gear) drives a much larger gear (the driven gear). The car moves slowly at first, and it takes a very high torque (lots of rotary force—which you experience as the engine racing) to get it rolling.

In a typical car, the gear ratios are as follows:

GEAR	RATIO	
1st (lowest)	3.72	
2nd	2.04	
3rd	1.34	
4th	1.00	
5th (highest)	0.82	

The first gear ratio, 3.72, means that the gear that turns the wheels (the driven gear) has 3.72 times as many teeth as the gear that is turned by the power of the engine (the driving gear). This means that it takes 3.72 revolutions of the driving gear to force one complete revolution of the driven gear.

Notice how the ratios drop as you move out of first. In fourth gear, acceleration would be very slow since the gear ratio is one-to-one. And in fifth gear, which is "overdrive," you go fastest but have almost no ability to accelerate (increase your speed) because the driving gear is actually larger than the driven gear.

In first gear it takes a great many revolutions of the engine to move the car. It's revolutions of the engine that burn up gas. So, driving in "first" uses lots of gas. In fifth gear, in contrast, it takes relatively few revolutions to move the car, making it the most economical gear.

On a 10-speed bicycle, the highest gear consists of a large gear driving a much smaller one. In this gear it is very hard to pedal, but each rotation of the pedal forces many rotations of the wheel so you will go much faster—if you have the strength!

Section 3: About Speed: Faster than a Speeding Bullet

Many things go fast. Did you know that rockets travel faster than a speeding bullet? Light travels at 186,000 miles per second? A rocket must achieve a velocity of 17,500 miles per hour to go into orbit around the earth? These speeds make driving a car at 55 miles per hour seem like barely crawling along the ground.

Miles per hour (mph) is a typical measure of speed. Traveling at a speed of 55 miles per hour means that if this speed is maintained, a distance of 55 miles will be covered in one hour. When taking a long trip, figuring out your average speed of travel makes driving less boring and lets you better plan your stops.

Here's how

AVERAGE SPEED—MPH

There's a very easy formula for computing average speed:

$$\text{Average speed} = \frac{\text{Total distance covered}}{\text{Total time}}$$

EXAMPLE: Fast Fred drove from Denver to New York in 34 hours (not counting rest stops). He covered a distance of 2,100 miles. What was Fast Fred's average speed?
SOLUTION:

$$\text{Average speed} = \frac{2,100 \text{ miles}}{34 \text{ hours}}$$
$$= 61.8 \text{ miles per hour (mph)}$$

Fast Fred traveled at an average speed of 61.8 miles per hour. To maintain this average, especially over as long a time period as 34 hours, means that Fast Fred probably drove for long stretches at very high speeds, perhaps even in excess of 80 mph—not the speed of sound certainly, but very, very fast. . . .

Do you know why you see the flash of lightning before hearing the accompanying roll of thunder?

Here's how

MACH AND SONIC BOOM

Lightning "comes first" because light travels many times faster than sound. Light travels at about 186,000 miles per second—there is nothing in the universe faster—while sound travels through *air* (at 32°F or 0°C) at about 742 miles per hour. The speed of sound in *water* at 32°F is about 3,246 mph—much faster than in air. (The speed of the propagation of sound is not constant: it depends upon the medium in which it travels and upon the temperature of the medium.)

Other things travel fast too. Commercial jet passenger planes travel at about 600 mph, while the Concorde travels at approximately 1,300 mph. The Concorde is called a *supersonic transport (SST)* because it travels faster than the speed of sound. When the SST travels through air at 32°F, it's going at the speed of Mach 1.8.

Named after the Australian philosopher and physicist, Ernst Mach, 1838–1916, *the Mach number is the ratio of the speed of an object to the speed of sound in the medium through which the object is traveling.* Therefore, the Concorde's Mach speed was obtained by dividing its speed by the speed of sound through air:

$$= 1,300 \text{ mph} \div 742 \text{ mph}$$
$$= 1.8$$

This means that the SST travels 1.8 times the speed of sound! As the speed of an airplane or rocket reaches and exceeds the speed of sound, a shock wave is created that is heard on the ground as a *sonic boom*.

Imagine a rocket traveling at 25,000 miles per hour. That's at about Mach 34 (25,000 mph ÷ 742 mph), or 34 times the speed of sound. The more you think about speeds, the more like science fiction it seems.

Here's how

ROCKETS . . . AND SPEEDING BULLETS

Rockets must achieve a very high speed (much faster than a speeding bullet) in order to break free of the pull of the earth's gravity. If a rocket reaches a speed of 17,500 miles per hour, it will go into orbit around the earth. If it travels still faster (25,000 miles per hour, to be precise) it will break completely free of earth's gravity and head off into outer space. The speed that must be reached at the surface of a planet to escape the pull of the planet's gravity is called the *escape velocity*.

Venus is a smaller planet than earth and the force of gravity at its surface is less than on our planet. To escape from Venus's gravitational pull requires a velocity of a mere 22,883 mph. Jupiter, on the other hand, is far bigger than earth and exerts a much greater gravitational pull on objects on its surface. The escape velocity on Jupiter is 134,548 mph. These large numbers may be a bit easier to imagine if we consider a comparison.

The muzzle velocity of a bullet fired from a .22 caliber rifle is 1,300 feet per second. This means that the bullet travels about one-quarter mile per second. Since there are 3,600 seconds in one hour, 1,300 feet per second is the same as $1,300 \times 3,600 = 4,680,000$ feet per hour. To translate this into miles per hour, we divide by 5,280 because there are 5,280 feet in a mile. Doing the division, we find that a .22 caliber bullet travels at the speed of 886.4 miles per hour (mph):

$$4,680,000 \div 5,280 = 886.4 \text{ mph}$$

The SST travels faster than a rifle bullet.

And a rocket headed toward the moon will reach a velocity of 25,000 miles per hour, or 28 times the speed of the bullet.

These speeds pale in comparison to the speed of light. Light travels at an incredible 186,000 miles per second (186,284 miles per second, to be precise.) Yes, that's *per second*. At this speed it takes light about one-seventh of a second to travel completely around the earth.

Since there are 3,600 seconds in one hour, light travels $3,600 \times 186,000$ miles $= 669,600,000$ miles in one hour. In one year, light would travel about 5.87 trillion (5,870,000,000,000) miles. The term "light year" refers to the distance light travels in one year. So, a light year is the same as about 5.87 trillion miles.

It takes light about one second to come here from the moon and about 8 minutes to travel from the sun to the earth. After the sun, the star nearest to the earth is Proxima Centauri, about 4 light years away. That is, it would take 4 light years to travel here from that star—a distance of about 24 trillion miles. And it would take our poor little rocketship, which travels at a mere 25,000 mph, 960,000,000 hours (about 110,000 years) to get to Proxima Centauri.

A decade ago, there were already plans for a new plane that could make the New York–Los Angeles trip in 12 minutes. It would travel at Mach 21!

Superman, watch out!

8

Hobbies,
Games, and Gambling

Section 1: Getting the Proper Exposure: Shutter Speed and Lens Opening

In 1898, when the first Kodak camera was marketed, it bore the slogan, "You press the button, we do the rest." All *you* had to do was line up the image, uncover the lens to admit light, press a button, and you had a picture.

As time went on, new discoveries in film and manual adjustments for focus, lens opening, and shutter speed gave the photographer more control over the final picture image. Today, many cameras—from the Kodak disc to the fancy foreign imports—are fully automatic. They have come full circle. The sophisticated developments in photography that occured in the last 100 years have all been built *into* the film and the camera. All you have to do is press the button, and the camera does the rest.

Since many of today's high-quality cameras have both an automatic and manual mode, however, they give you the option of doing more than that. Typically, what they let you do (or in some models, require that you do) is:

- *Focus.* (There are several different, easy methods, some involving viewfinders or rangefinders, that are described in the directions that come with the camera.)
- *Control the light* (or exposure—more about this below).

The result is that you can take pictures under a wide variety of conditions and situations.

Your purpose in controlling the light entering the camera is to assure that the right amount comes through the lens onto the film. Too much light and the picture comes out too light (overexposed). Too little light and the picture is too dark (underexposed). When the needle on your light meter (don't forget to turn it on) balances at the center of the scale, you have the right amount of light.

There are two manual adjustments for regulating the available light: you can vary the speed of the shutter and/or change the size (width) of the lens opening.

Here's how

SHUTTER SPEED

Simply stated, the shutter is a highly crafted lid inside the camera that covers the film, barring light. When you set the shutter speed, you determine how long the film will be uncovered and, as a result, how much light will pass through the lens and strike the light-sensitive particles on the film.

Shutter speed is measured in *fractions* of a second and is noted on your camera as 500, 250, 125, 60 (and maybe 30, 16 . . . and B—for bulb). By tradition, camera manufacturers have left off the top of the fractions, so that you have to interpret them. When you set the shutter to one of those numbers just noted, you are uncovering the film for exactly 1/500th, 1/250th, 1/125th, or 1/60th of a second.

Remember fractions? You might recall, for example, that 1/60th is a larger piece of a pie than 1/250th (just as 1/2 is more than 1/4, and 1/4 is larger than 1/8). Thus, at a shutter speed of 60, the shutter is open longer than at 125, 250, or 500. The longer the lens is open, the more light can reach the film.

The scale goes like this:

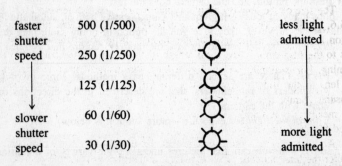

faster shutter speed	500 (1/500)		less light admitted
	250 (1/250)		
↓	125 (1/125)		↓
slower shutter speed	60 (1/60)		more light admitted
	30 (1/30)		

All other things being equal, under average light conditions, a shutter speed of 125 is the usual setting. When the light is exceptionally bright, however (as it is at the beach on a sunny summer day), you must keep the shutter open a shorter time (use a faster speed—say 250 or 500) to admit the right amount of light. On a dark, cloudy day, or whenever there is not a lot of available light, the reverse is true: keep the shutter uncovered for a relatively long time—otherwise the picture will be too dark.

At shutter speeds of 60 or slower, it becomes difficult to hold the camera still enough for the entire fraction of a second so that the image doesn't blur from your own motion. (That's why the tripod was invented.)

✓*TIP* If you don't have a tripod, use your own body as one: lean against a solid object, press your elbows into your sides, take a deep breath and hold it while shooting.

To capture a sharp, unfuzzy image of a moving object (whether it be a runner or leaves on a tree waving in the breeze), it is necessary to use a very fast shutter speed, such as 250 or faster. This can only be done if there is good light available. When the light isn't bright enough, the fast shutter speed probably won't admit enough light. And, if you slow down the shutter speed to let in the amount of light you need, you will lose the sharp image. Can you take a picture under these conditions? Probably yes, but you need to make a further adjustment for light.

Here's how

LENS OPENING

By varying the *width* of your lens opening (aperture—the hole the light goes through), you have added control over the amount of light hitting the film. A small opening lets through less light than a large one.

The size of the lens opening is typically noted on the camera as 2, 2.8, 4, 5.6, 8, 11, 16. These numbers are called f/stops. (Now, the photographic jargon.) An f/stop is the ratio of the distance from about the center of the lens to the film (this is the focal length of the lens) to the size of the lens opening. For example, with a 50mm lens, the distance from the center of the lens to the film is about 50mm. If the diameter of the lens opening measures 25mm, then the f/stop would be 2.0 or $50mm \div 25mm = 2.0$. What this means is that smaller lens openings give larger ratios, or higher-numbered f/stops (and vice versa).

Another way of saying this is that the higher the f/stop number, the smaller the lens opening. This is what it looks like on a simple chart:

smaller ratio	f/stop	size of lens opening		more light admitted
	2			
	2.8			
	4			
	5.6			
	8			
	11			
larger ratio	16			less light admitted

(If you look through the lens of your camera when the back is open and there's no film in it, you can actually see the opening at the back of the lens enlarge and shrink in diameter as you go back and forth from low to high f/stops.)

Why use a particular f/stop? Why not just pick the best shutter speed for the kind of object you're photographing (stationary, moving) and for the available light conditions (bright, dim) and then fiddle around with the f/stop adjustment until your light meter balances?

With higher f/stops, the narrow lens opening not only admits little light, but also has the effect of focusing the incoming light in a relatively narrow beam. This gives a clear image of objects *over a distance*. To test this, look at an object 10 to 20 feet away. Now squint your eyes (narrowing them) and see how much clearer the object becomes.

✓✓*HOT TIP* If you've left your reading glasses at home, squinting your eyes and holding the material farther away from you will make it more legible for short periods. On the same principle, looking at an object through a pinhole in a piece of paper has the effect of focusing the image more clearly.

The improved focus resulting from a small lens opening (high f/stop) is referred to as "greater depth of field." With a larger lens opening, you can take a picture with less available light, but the trade-off is a *shallow* depth of field in which the only objects in focus in your final picture will be those you focused on—not the surrounding foreground or background.

Here's how

GETTING THE RIGHT BALANCE

What combination of shutter speed and lens opening will give the needed amount of light for highlight and shadow detail and also provide a clear, sharp image of details in the objects in the foreground and background?

No one *single* combination works because of the mathematical relationship between the two scales. For example, if the two adjustments (shutter speed and width of aperture or f/stop) are made in equal numbers of steps in opposite directions, there is no change in the light coming into the camera. All of the shutter speed/lens opening pairings below admit the same amount of light:

Shutter Speed		f/Stop		
500	fast speed	4.0		wide opening
250		5.6		
125		8.0		
60	↓	11.0		↓
30	slow speed	16.0		narrow opening

Notice how many different combinations there are. Also, take note of the relationships:

- The lens opening decreases as the shutter speed decreases
- The slower the shutter speed, the higher the f/stop
- The smaller the f/stop, the faster the shutter speed

Using your camera, experiment with various combinations under different lighting conditions. Be particularly mindful of the changes in depth of field in your final picture.

Here are some ideas for adjusting your camera for the proper exposure. Let's assume you have lots of light to work with. Unless the object you're photographing is in fast motion, start out by setting your shutter speed at 125. Look through the lens and adjust your f/stop until the light meter balances. OK? Take a deep breath, hold it, and . . .

Shoot!

If you're trying to get a sharp picture of, let's say, a racer under these same light conditions, you will want a faster shutter speed, say, 500. After you set this speed, see if there is an f/stop that will balance the needle; if not, you may be forced to use a slower speed, sacrificing some sharpness of image.

To find the fastest possible speed, open the lens as wide as possible (the smallest f/stop on your camera), look through the lens at the light meter, and adjust the shutter speed until the needle moves past the halfway mark. Then make a final small adjustment in the lens opening to bring the needle into balance.

Let's take this further and suppose that depth of field is important in this picture. You want as much as possible, near and far, to be in focus. So, start by setting the f/stop at say, 11, and then manipulate the shutter speed until the light meter is in balance or comes close to balancing. (If it comes close, you can bring it into more exact balance by slightly adjusting the f/stop.)

Up to now, we've been assuming normal light. If the light is dim, you may have to forgo great depth of field as you open your lens as wide as possible.

Light control is an important part of the technique that distinguishes an Alfred Stieglitz from the Sunday shutterbug. Manually setting the lens opening/shutter speed combination allows you to supersede external conditions of light and motion and gives you more say in the kind of image you want. You will be less limited in your picture-taking, and your new skills will help satisfy your secret belief that you are indeed an exceptional photographer.

Section 2: Playing the Odds: An Introduction to Probability

If knowing the odds was all it took to win at backgammon, poker, blackjack, or craps, then anyone could become a mathematics major specializing in probability and statistics, go to Las Vegas, Reno, Atlantic City, or Monte Carlo, and become rich—provided that the casinos did not ask you politely, but firmly, to leave because of your dazzling skill and phenomenal winnings.

But it takes more than an understanding of probability to win at games of chance. Winning also requires strategy, the ability to bluff, luck, and many other factors that are not quantifiable. Indeed, knowing the odds may lead you to conclude that you don't stand a chance.

While understanding probabilities won't assure your success as a gambler, most professional gamblers and dedicated amateurs are certainly keenly aware of probabilities, at least in an intuitive way. Experienced backgammon or craps players, for example, know the possibilities of various possible outcomes of rolls of the dice.

Here, we provide a brief introduction to the concept of probability and odds. In the next section, we discuss some of the important probabilities for backgammon, poker, and craps. We also present tables of some of these probabilities, for, while we don't encourage memorization as the way to master mathematics, we certainly don't expect you to interrupt an exciting card game in order to carefully compute the probabilities before deciding on your next move. If you did this, your opponent would have a fit—justifiably—if she didn't walk away instead.

What are probabilities and how are they determined?

Here's how

<hr>

PROBABILITY

Ask yourself what is the *probability* (in other words, the likelihood) of getting a head when flipping a fair (meaning a two-sided and well-balanced) coin one time? If you answered 50%, even, 50–50, ½, or 0.5, you would be right to some degree, although, strictly speaking, only the last two answers (½, 0.5) are correct. Let's see why.

Flipping a coin is an example of a *random experiment*—an act in which the outcome is determined entirely by chance. With a fair coin, the outcomes of "getting a head" or "getting a tail" are equally likely. *To determine the probability of getting either one, we compute the ratio of number of favorable outcomes to the number of possible outcomes:*

$$\text{Probability} = \frac{\text{Number of favorable outcomes}}{\text{Number of possible outcomes}}$$

In flipping a coin, the number of possible outcomes is 2 (heads or tails). The number of favorable outcomes (by which we mean the number of outcomes that is of interest to us), let's say heads, is 1. Therefore:

$$\text{Probability of a head} = \frac{1}{2}$$

(Notice that the probability of getting a tail is also ½). We could write ½ as 0.5 and also say the probability of getting a head is 0.5, as is the probability of getting a tail.)

In a random experiment in which all outcomes are equally likely, *the probability of an event (an outcome or set of outcomes) is the ratio of favorable to possible outcomes*. In this kind of situation, the following facts about probability are important:

1. P (Probability) = 0 means the desired (favorable) outcome is impossible. There are no favorable outcomes.
2. P = 1 means the desired (favorable) outcome is certain. For example, the probability of getting a head *or* a tail when flipping a coin once is one, because the number of favorable outcomes is 2 (heads *or* tails) and the number of possible outcomes is also 2:

$$\frac{2}{2} = 1.$$

Thus, the ratio of favorable outcomes to possible outcomes is 1.
3. If P is the probability of a particular event, then $0 \leq P \leq 1$ (P is greater than or equal to zero and less than or equal to 1.) In other words, the value of P falls between 0 and 1.

Now let's look at some examples of probabilities involving dice.

A die (the singular of dice) is a cube with 6 faces, each of which is imprinted with between 1 and 6 dots corresponding to the numbers 1 to 6. When you roll one die, there are exactly six possible outcomes: 1, 2, 3, 4, 5, or 6. Let's figure out the probability of rolling a 4. Here there is one favorable outcome and 6 possible outcomes. So the probability of getting a 4 is ⅙.

Now, what is the probability of rolling either 2 or a 3? In this case, there are 2 favorable outcomes (2 *or* 3) and 6 possible outcomes. Thus, the probability of rolling a 2 *or* 3 is ⅔ or ⅓.

Probabilities become more difficult to compute as the number of favorable and possible outcomes become more difficult to count. We'll do a few more examples with dice to give you an idea of what this means.

In most games involving dice, you roll a *pair* of dice. Before we begin computing the probabilities of various outcomes, we need to count the number of possible outcomes. To do this, it is important to understand that even though the two dice may look alike, there are really two distinct dice being rolled. This is most easily imagined by thinking of the pair as consisting of one red die (R) and one green die (G). Now, we'll count all possible outcomes by listing all R–G combinations:

The first column shows that when we roll 1 with the R die, we can roll from 1 to 6 with the G die. The second column shows rolling a 2 with the R die and 1 to 6 with the G die. The rest of the columns are read the same way.

Table 1
Possible Outcomes of Rolling Two Dice (R and G)

R	G	R	G	R	G	R	G	R	G	R	G
1	1	2	1	3	1	4	1	5	1	6	1
1	2	2	2	3	2	4	2	5	2	6	2
1	3	2	3	3	3	4	3	5	3	6	3
1	4	2	4	3	4	4	4	5	4	6	4
1	5	2	5	3	5	4	5	5	5	6	5
1	6	2	6	3	6	4	6	5	6	6	6

Getting a 2, 3 means getting a 2 on the R die and a 3 on the G die. This is not the same as getting 3, 2 or a 3 on the R and a 2 on the G die, although the sum, 5, is the same in both cases.

Altogether, we have listed a total of *36 possible outcomes*.

Now for some examples.

EXAMPLE: First, what is the probability of getting a sum of 5 when rolling a pair of dice?

SOLUTION: To answer this, we are required to count the number of favorable outcomes (keeping in mind that there are 36 possible outcomes). That's the number of different outcomes that result in a sum of 5.

We could get a five in each of the following four ways:

1,4   4,1   2,3   3,2

Therefore:

$$\text{Probability (sum} = 5) = \frac{\text{Number of favorable outcomes}}{\text{Number of possible outcomes}} = \frac{4}{36} = \frac{1}{9}$$

EXAMPLE: In this next example, we want to know the probability of rolling a 7 (that is, getting a sum of 7) using a pair of dice.

SOLUTION: To solve this problem, we first list the number of favorable outcomes:

1,6   6,1   2,5   5,2   4,3   3,4

Since there are 6 favorable outcomes and 36 possible outcomes, the probability of rolling a 7 is $\frac{6}{36}$ or $\frac{1}{6}$.

The probability of getting every other sum from 2 through 12 can be computed in the manner we just described. The following is a list of the probability of each sum when rolling a pair of dice:

SUM	PROBABILITY
2	$1/36$
3	$2/36 = 1/18$
4	$3/36 = 1/12$
5	$4/36 = 1/9$
6	$5/36$
7	$6/36 = 1/6$
8	$5/36$
9	$4/36 = 1/9$
10	$3/36 = 1/12$
11	$2/36 = 1/18$
12	$1/36$

Note that the most difficult outcomes to roll are a sum of 12 or a sum of 2. That's because there is only one way to get a 12 (6,6) or to get a 2 (1,1). In contrast, a sum of 7 has the highest probability, $6/36$ or $1/6$, because there are more ways to get a sum of 7 than there are ways to get any other sum. You might want to see if you can list all the ways to get a few of the other sums.

Odds are related to but a bit different from probability. The concept of "odds" is an integral part of betting.

Here's how

AGAINST ALL ODDS

If you were asked to bet on the outcome of the flip of a coin, you would be likely to offer "even money." That is, you might bet $1 on heads against your opponent's $1 on tails. Intuitively (and quite correctly), you are likely to think of odds as the *ratio of the number of outcomes favorable to your bet to the number of outcomes that are not favorable to your bet.* When flipping a coin, for example, the odds in favor of a bet on heads against tails are 1 to 1. This is commonly phrased as "50–50." Notice that the ratio 50 to 50 is equivalent to the ratio 1 to 1.

Now think about rolling a pair of dice again. We already figured out that the probability of getting a sum of 7 was $6/36$ or $1/6$. The odds in favor of getting a sum of 7 is the ratio of the number of outcomes that are favorable to this result (namely, 6) to the number of outcomes that are not favorable. The unfavorable outcomes are *all* the outcomes that do not result in 7.

Since there are 36 possible outcomes, that's 36 minus 6 or 30 outcomes that are not favorable. So the odds in favor of getting a sum of 7 are 6 to 30 or 1 to 5. Clearly, the odds are not in your favor (although they are better than for any other sum). If you wanted to bet on getting a sum of 7, you should bet $1 against $5.

A horse named Turtle (not a good omen!) is a "20 to 1 shot." In horse racing, this means that the odds in favor of him *losing* are 20 to 1. In this case, a fair bet would be your $1 in favor of Turtle's winning to someone else's $20 bet against him. In probabalistic terms, out of every 21 races, you should expect poor Turtle to win only once and lose 20 times. Thus, the probability of his winning is ½₁.

If you know the probability, you can always compute the odds.

If the probability of an event is $\frac{a}{b}$, the odds in favor of the event are a to (b − a). For example, when rolling a pair of dice, the probability of the event "snake eyes" (that's 1,1) is ⅟₃₆. The odds *in favor* of snake eyes are a to (b − a) or 1 to 36 − 1 = 1 to 35. (The odds *against* snake eyes are 35 to 1.)

As another example, suppose that the feeling is that the chances of the New York Mets winning the pennant are about 7 out of 10 because of their new players. That's the same as ⁷⁄₁₀. Then, according to this estimate, the *odds in favor* of the Mets would be 7 to 3 (or 7 to [10 − 7]). The odds against them would be 3 to 7. A fair bet in *favor* of the Mets would be $7 against $3, or $14 to $6 and so on in a 7 to 3 ratio.

Now, you may feel ready to place a bet. We hope not. There's no such thing as a sure thing. Remember, if there's an 80% chance that you'll win, there's also a 20% chance that you'll lose. That's 1 out of every 5 chances against you. We hope that knowing the odds will make you aware of those against you as well as those in your favor.

Section 3: Games of Chance: Some Facts about Backgammon, Poker, and Craps

Now that you've read Section 2 and know the principles of computing probabilities, you're ready to consider some probabilities that are basic to the games of backgammon, poker, and craps. If you don't know these games, parts of this section will not make much sense, although you may still find the discussion of probabilities interesting.

We'll start with backgammon.

Here's how

BACKGAMMON: "THE BLOT"

In the game of backgammon, the number of moves of your markers is controlled by the outcome of rolling a pair of dice. Here, however, it is not the total sum that counts, but the numbers appearing on each of the "up" faces of the two dice. For example, a roll whose outcome is 5,3 enables you to move one marker 5 and the same or another marker 3 spaces. The choice of which marker to move how many spaces is not always up to you. If, for example, your opponent has made a point (has two or more markers on a position), you are not permitted to terminate a move at that position. Thus, you may not be able to move, say, one of your markers 3 spaces because the point you would land on has been made by the other player.

On the other hand, you can be "hit" (bumped off the board) if your opponent lands a marker on a position on which you have only one marker. Such a position is called a "blot." Let's consider the probability of a blot being hit. To simplify the computation, we will assume that your opponent is free to land anywhere before your blot.

If your blot is 5 points (positions) away from one of your opponent's markers, what's the probability that you'll be hit? A hit requires that one of the numbers appearing on the face of the dice is a 5 or that the sum of numbers that appear is 5. Here's a list of all the ways that a 5 can appear or that the sum of the two dice will total 5:

5 APPEARS:		SUM = 5:	
5,1	1,5	1,4	4,1
5,2	2,5	2,3	3,2
5,3	3,5		
5,4	4,5		
5,5			
5,6	6,5		

Counting, there are 11 ways for a 5 to appear and 4 ways for the sum to be 5. Therefore, your blot can be hit in 15 (11 + 4) ways. The probability of being hit is $^{15}/_{36}$ (36 is the number of possible outcomes of throwing the dice). The fraction $^{15}/_{36}$ is equivalent to about 42%. This means that if your blot is 5 points away from the other player's marker, you can expect it to be hit about 42% of the time.

Now, suppose that your blot is 7 units away from your opponent. The chance of being hit becomes much slimmer, because you cannot be hit by a 7 appearing directly, but only by the sum of the outcomes of the roll of the two dice. We saw in the last section that there are 6 ways the two dice can total 7:

| 1,6 | 6,1 | 2,5 | 5,2 | 3,4 | 4,3 |

The probability is $\frac{6}{36} = \frac{1}{6} = 17\%$. The dramatic drop in the likelihood of being hit is due entirely to the fact that there is no way for a 7 to appear on one die only, since a die has only 6 faces.

Below is a table of all the relevant information on the probability of a blot being hit:

Table 1
Probability of "Blot" Being Hit in Backgammon

DISTANCE AWAY	WAYS TO BE HIT	PROBABILITY OF A HIT	PROBABILITY AS A %
1	11	11/36	31%
2	12	12/36	33%
3	14*	14/36	39%
4	15*	15/36	42%
5	15	15/36	42%
6	17*	17/36	47%
7	6	6/36	17%
8	6*	6/36	17%
9	5*	5/36	14%
10	3	3/36	8%
11	2	2/36	6%
12	3*	3/36	8%

*These outcomes can be reached in extra ways by getting doubles.

In backgammon, when you roll doubles, you can move the amount you get twice. For example, double 4's means you can move 4 and 4, and then 4 and 4 again. This can be done with either the same marker or with different markers. If your opponent got double 4's, he or she could possibly hit your blot 12 points away by moving one marker 4, 4, 4. In fact, a blot 16 points away would also be vulnerable to a hit when double fours are rolled (4, 4, 4, 4). However, the probability of rolling double 4's is only $\frac{1}{36}$, or 3%.

It's worth examining the figures in Table 1 a little more closely. Instinctively, you may feel most at risk when your blot is closest to your opponent, but the figures show this is only partly true. If your blot is more than 12 points away, it is either very unlikely or impossible to be hit. And, if you look at the table, you'll see it's safer to be 10 points away than 5 points away. But notice that 12 is less safe than 11 and no more safe than 10. And it is much safer to be 1 point away than it is to be 6 points away.

✓*TIP*   One conclusion to be drawn from these probabilities is that if you must expose a blot less than 7 points away from your opponent, then the closer the better!

There's another important situation in backgammon that involves probabilities. This occurs when you've been hit and knocked off the board onto the bar.

Here's how

BACKGAMMON: ENTERING FROM THE BAR

Rolling a number between 1 and 6 enables you to land on points 1–6 on your opponent's inner table and thus return your marker to the game. This becomes more difficult if your opponent has made some of the points on his inner table. Then the number of points you can enter on is reduced accordingly.

If all 6 points are open, you are sure to reenter the game on the next roll of the dice. Suppose, however, that your opponent has made one of the points on the inner table—say, point 5. In that case, you will be able to reenter the board with any throw *except* double 5's. So the probability of reentering would be $^{35}/_{36} = 97\%$. If your opponent has made 2 points, 4 and 5 for example, you would be able to reenter with any throw of the dice except:

4,4   5,5   4,5   5,4

There are only 4 ways you can't enter, so there are 32 ways (36−4) in which you *can* enter. Thus, the probability of returning from the bar when the other player has made two points is $^{32}/_{36} = 89\%$. The following table (Table 2) shows the probabilities of entering from the bar, varying by the number of points your opponent has made:

✓*TIP*   Looking at the probabilities, you can be pretty sure of being able to return your man to play after being hit if your opponent has made 1 or 2 points in his inner table. If he has made 3 points, there is a 25% (100%−75%) chance that you will be unable to reenter on a given throw of the dice and will, therefore, lose your turn. Things look increasingly bad after this, so you might become more careful about exposing blots as your opponent gets to the position of having 3 or more points in his inner table.

In the last section we pointed out that in real play you can't stop to compute probabilities. Instead, we suggested that you memorize some of the

Table 2
Probability of "Entering from the Bar" in Backgammon

NUMBER OF POINTS MADE BY OPPONENT	NUMBER OF WAYS TO COME IN FROM THE BAR	PROBABILITY OF ENTERING	PROBABILITY AS A %
1	35	35/36	97%
2	32	32/36	89%
3	27	27/36	75%
4	20	20/36	56%
5	11	11/36	31%

probabilities we illustrated so that you will have them at your disposal while you play. You might also improve your game by reading through any of several good books that offer extensive discussions of backgammon probabilities and strategy.

Good luck! Here's hoping you win . . .

Craps is *not* a dirty word! It's the name of a popular gambling game that developed from an old English game called Hazard. It was introduced into the United States in the early nineteenth century.

Here's how

SHOOTING CRAPS

Craps is usually played as follows. The shooter (of a pair of dice) keeps on rolling the dice until she either wins or loses. There are several different ways to win or lose.

1. Rolling a sum of 7 or 11 on the first roll wins.
2. Rolling a sum of 2, 3, or 12 on the first roll loses.
3. Rolling any other sum (4, 5, 6, 8, 9, 10) on the first roll means you continue to roll the dice until you either match your first sum or roll a 7.
4. Matching your first sum before rolling a 7 wins.
5. Rolling a 7 before matching your first sum loses.

The probability of winning at craps is not easy to compute. But the basic game just described is very fair. In fact, under these rules, the probability of winning is almost ½. It's actually about 0.493 as we show you on the next page. That is, you would have about a 49.3% chance of winning.

In casinos, the rules may change slightly, resulting in changes in the probabilities. (You can be sure the change is in the casino's favor, not yours.)

Let's figure out the probability of winning and the probability of losing on the first roll of the dice. To win, you must roll a sum of 7 or 11. Earlier in this chapter (Section 2), we found that there were 6 ways to roll a sum of 7 and two ways to roll a sum of 11. Thus, the probability of rolling a 7 or 11 is $(6+2)$ $\frac{8}{36}=0.22$. You have about a 22% chance of winning on your first roll.

You'll lose if your first roll is a 2, 3, or 12. There is one way to roll a sum of 2, two ways to roll a sum of 3, and one way to roll a sum of 12. Therefore, the probability of rolling a 2 or 3 or 12 is $\frac{4}{36}=.11$. So there's about an 11% chance that you will lose on your first roll.

The difficult computation is in figuring out the probability of winning *after* rolling a 4, 5, 6, 8, 9, or 10. You can roll any number of times *as long as you don't roll a 7!* We have figured out these probabilities for you and listed them below:

Table 3
Winning Probabilities in Craps

IF YOUR FIRST ROLL IS:	THE PROBABILITY THAT YOU'LL MATCH NUMBER BEFORE GETTING A 7 IS:
4	$1/3 = 33\%$
5	$2/5 = 40\%$
6	$5/11 = 45\%$
8	$5/11 = 45\%$
9	$2/5 = 40\%$
10	$1/3 = 33\%$

Because there are more ways to make 6's and 8's, you stand a better chance of matching those numbers before rolling a 7 than you do if your "point" is 4 or 10. The adventurous reader might check that the probability of winning at craps, 0.493, is obtained by computing the following:

$P(4) \times \frac{1}{3} + P(5) \times \frac{2}{5} + P(6) \times \frac{5}{11} + P(8) \times \frac{5}{11} + P(9) \times \frac{2}{5} + P(10) \times \frac{1}{3} + P(7 \text{ or } 11) \times 1$

Here, for example, $P(4)$ stands for the probability of rolling a 4 and $\frac{1}{3}$ is the probability of matching that 4 before rolling a 7 as given in Table 3 above. The values for $P(4)$, $P(5)$, etc. are given in the table on page 172 of Chapter 8, Section 2.

This is just a sample of the most basic probabilities involved in the basic game. The game gets more complicated with side bets and different combinations of points.

Craps can be fun *and* you have a reasonable chance of winning, yet we still don't recommend gambling. After all, you also have a more than reasonable chance of losing.

Poker is one of the classic card games. It is also a gambling game conjuring up images of a windowless back room filled with heavy aroma of cigar smoke, a round table piled high with chips, and a group of serious, professional gamblers.

Here, we take a more mathematical view of the game.

Here's how

PLAYING FIVE-CARD POKER

In five-card poker (one of the many variations of the game), each player is dealt 5 cards from a deck of 52. The object is to get groupings of cards that form pairs (two of the same numbered cards), 3 of a kind (three of the same numbered cards), 4 of a kind (four of the same numbered kind), a straight (five cards in numerical order), a flush (all five cards of the same suit), a full house (three of a kind and a pair) and, best of all, a straight flush (all five cards of the same suit in numerical order). The ideal is the famous royal flush—from 10 to the ace in the same suit.

Computation of probabilities in poker requires an understanding of permutations and combinations. This is a complicated topic that is best learned in a college course in finite mathematics or statistics and probabilities. Here we will just show you some of the results and offer you some tips.

The difficulty in calculating probabilities in poker comes from the large number of possibilities that exist. We just can't list them all as we did with the 36 possible outcomes with dice. The number of possible poker hands (5 cards out of 52) is enormous—2,598,960!

The probability of being dealt a pair is 0.42. That's a pretty good chance. On the other hand, the probability of being dealt four of a kind from a full deck is 0.00024. Hands with the lowest probability are the most valuable—because they are the most rare—and always beat hands with higher probabilities.

Below, we have listed the probability of being dealt various hands:

Table 4
Probability of Types of Poker Hands

TYPE OF HAND	NUMBER OF WAYS	PROBABILITY
One pair	1,098,240	0.42
2 pair	123,552	0.05
3 of a kind	54,912	0.02
Straight*	10,200	0.004
Flush*	5,108	0.002

Table 4 (*continued*)
Probability of Types of Poker Hands

TYPE OF HAND	NUMBER OF WAYS	PROBABILITY
Full house	3,744	0.001
4 of a kind	624	0.0002
Straight flush	40	0.00002

*Straight flushes are not included in these figures.

You can see from the table that while being dealt a pair is fairly likely, straight flushes are truly rare occurrences, happening in only 2 out of 100,000 chances. And of the 40 ways to be dealt a straight flush, only 4 of them are royal flushes.

These probabilities are just the rudiments of what you have to know before you can play the game of poker effectively. As we've said before, in poker as in all games of chance, just knowing the basic rules and probabilities are not enough—although they are very important. *Experienced* gamblers also have well thought-out strategies, know when to bluff, and understand the value of luck. Most important, *successful* gamblers know to quit when they're ahead.

PART THREE

INDOOR MATH

9

Media
Math

Section 1: Dealing with Price Fluctuations:
The Consumer Price Index

Suppose for a moment that apples sold at an average price of 20¢ last year
but cost an average of 25¢ apiece today. To answer the question, "By what
percentage has the price of applies increased?", we calculate the percent
increase (see Chapter 1, Section 1.) We do this by subtracting last year's
price from today's price, dividing the answer by last year's price, and multiplying by 100:

$$25 - 20 = 5$$
$$5 \div 20 = 0.25$$
$$0.25 \times 100 = 25\%$$

Another way of looking at the change in price is to consider the old price
as 100% of itself. Then the new price is 125% (100% + 25%) of the old
price. These computations answer the question, "What is the cost today of
some amount of goods and services compared to the same amount in some
base period year?"

Assuming the apple represents *all* the goods and services we are interested in, last year can be considered the base year in which the price is
100%. This year's price (125%) is called an *index number*. The 125 simply
says that this year's price is 125% of the base year's price. Since the base
year price is 100(%), this year's price is 25% higher.

The *Consumer Price Index* (CPI) is a monthly economic index number of the Bureau of Labor Statistics that measures year-to-year *changes in the average price of goods and services*. In its present form, the CPI dates back to World War I, when very rapid rises in prices, especially in shipbuilding centers heavily involved in the war effort, made such an index essential for calculating cost-of-living adjustments in wages. To this day, the CPI is used to determine wage policy in all kinds of labor negotiations, as well as calculating adjustments to Social Security payments and in other situations (e.g., food stamps, the school lunch program) where cost-of-living changes need to be considered to maintain the real purchasing power of the dollar.

In addition to its usefulness as an indicator of change in the cost of living, the CPI is also a measure of inflation/deflation in the economy and is valuable as a means of studying trends in prices of various goods and services. In these ways, the CPI can be taken as a measure of the effectiveness of national economic policy.

How did the CPI originate?

Here's how

INDEX NUMBERS

Historically, interest in economic indexes peaks at times of great price fluctuations. The earliest economic index was little more than the average of prices paid for grain, wine, and oil at two different time periods. The Napoleonic wars and other major events like the California and Australian gold discoveries stimulated new investigations of price fluctuations. In 1864, the economist, Étienne Laspeynes, developed index formulas that are the basis of those used today by the Bureau of Labor Statistics.

Defined by the Bureau of Labor Statistics as a measure of price change, *the Consumer Price Index is the ratio of the cost of a particular series of consumer items now to the cost of the same series of items at some set time in the past (the base or reference period).**

The current Bureau of Labor Statistics' formula is rather impressive. It is essentially unchanged since first developed by Laspeynes and looks like this:**

---

*This Section is based extensively on the Bureau of Labor Statistics' *Handbook of Methods, Volume II, The Consumer Price Index,* April 1984, U.S. Department of Labor, Bureau of Labor Statistics. The Bureau of Labor Statistics has been publishing economic indexes since 1919.

** Although the Bureau of Labor Statistics distinguishes between the base period for prices and for quantities, we have simplified the discussion by avoiding this distinction.

$$I_{t0} = \frac{P_{1t}Q_{10} + P_{2t}Q_{20} + \cdots P_{it}Q_{i0}}{P_{10}Q_{10} + P_{20}Q_{20} + \cdots P_{i0}Q_{i0}} \times 100$$

Before skipping ahead, let us define the symbols.

I_{t0} is the index number for t, the comparison period, and 0 is the base period. The letter i stands for the total number of items in the market basket. P_{1t}, P_{2t}, etc. are the prices for items 1, 2, etc. in the comparison (present) period t; P_{10}, P_{20}, etc, are the prices for the items in the base period 0; and Q_{10}, Q_{20}, etc. are the quantities of items 1, 2, etc. consumed in the expenditure base period.

$P_{1t}Q_{10}$ is the total cost of the first item in the market basket at time t because you multiple the price of the item by the quantity. The numerator of the I_{t0} formula is thus the total cost of the market basket at time t. $P_{10}Q_{10}$ is the total cost of the first item in the base period, so the denominator is the total cost of the same market basket at the base period. I_{t0} is the ratio of the cost of the market basket now its cost in the base period. (We multiply by 100 to convert the ratio to a percent.)

What, you may ask, are the series of items that are included in the fixed market basket of goods and services, and how are they priced?

Here's how

CPI

The CPI is based on a sample of actual prices of goods and services grouped under 7 major categories of consumer expenditures: food and beverages, shelter and fuels, apparel and upkeep, transportation, medical services, entertainment, and "other goods and services" that people buy for day-to-day living.

Each major group is subdivided. Take "food and beverages" as an example. The first division is "food at home—outside the home." Under "food at home" there are cereals and bakery products; meats, poultry, fish and eggs; dairy products, fruits and vegetables; and "other foods at home." Within "meats, poultry, fish, and eggs," for example, there is beef and veal, pork, other meats, poultry, fish and seafood, and eggs. A further subdivision of "pork," as an illustration, includes: bacon; pork chops; ham, other than canned; pork other than bacon, chops, ham, sausage; and canned ham. In total, there are 382 of these items that are priced for the CPI.

Price change, according to the Bureau of Labor Statistics, is measured by "repricing essentially the same market basket of [382] goods and services at regular time intervals [monthly] and comparing the aggregate costs with the costs of the same market basket [both the same items and the same *quantity* items] in a selected base period."

Beginning with the economic data for January 1988, the base period for the Consumer Price Index is 1982–84. Thus, the cost of today's market basket can be compared with the cost of the same goods and services for each year for the past 10 years. (Historical data on this new reference base are available from the Bureau of Labor Statistics for all currently published national and local area indexes.)

The CPI is continuously undergoing revision. The most recent major revision was completed in 1987. While it did not change the items in the market basket of goods and services, the new revision incorporated new "expenditure weights" (the "Q's" in the formula—that is, how much of a particular item was consumed by the population during 1984, the most recent study period). It also introduced new retail outlets (now prices are obtained from 21,000 retail establishments and 60,000 housing units in 85 urban areas across the country). Finally, the most recent revision added in new population data from the 1980 Census. It also improved sampling, data collection, processing, and statistical estimations. The relatively new CPI–U (1978) is supposed to reflect the "buying habits of approximately 80% of the non-institutionalized population of the United States." Previously, the CPI (CPI–W) represented the buying habits of only urban wage earners and clerical workers, which covered about half of all urban consumer groups in the United States.

Today, the Bureau of Labor Statistics publishes both the CPI–U and CPI–W. The CPI–W, however, is still the basis of cost-of-living adjustments in most wage negotiations.

The Consumer Price Indexes are also presented in "seasonally adjusted" and "seasonally unadjusted" ways. (Unless otherwise noted, the CPI is quoted for unadjusted data.) Seasonally adjusted data are often preferred for analyzing general price trends in the economy because they eliminate the effect of changes in buying patterns that occur at the same time every year—such as price movements resulting from holidays, sales, and model changeovers.

Consumer Price Indexes are also available for various geographic locations and specific segments of the total market basket: food and beverages at home, for example, or energy prices. It is not unusual to read or to hear on the news that "the New York–Northeastern New Jersey [one of the local areas] CPI went up 0.3% in April—3.6% from a year ago" or, "Grocery prices in the Buffalo metropolitan area up 1.3% in March—2.1% from a year ago."

How are the calculations done?

Here's how

CPI PERCENT CHANGE

The index measures price changes from the 1982–84 reference data, which equal 100(%). An increase of 125%, for example, is shown as 225. According to the Bureau of Labor Statistics, the price of a base period market basket of goods and services can also be expressed in dollars: from $10 in the base period to $22.50 today (225% of $10 = 2.25 × $10 = $22.50).

There are two ways the Bureau computes changes from one time period to another: *index point change* and *percent change.* "Movements of the indexes from one month to another are usually expressed in percent changes rather than changes in index points because index point changes are affected by the level of the index in relation to its base period while percent changes are not." The index number is always a percent of the *base period price* (1982–84). In contrast, a percent change can compare any two periods.

The following example reproduced from the May 1991 Bureau of Labor Statistic's *News* illustrates how index points and percent changes are calculated.

Index point change

CPI (in Period B) 115.7
Less previous index (CPI in Period A) 111.2
Equals index point change 4.5

Percent change

Index point difference (between Period A and B) 4.5
Divided by the previous index (Period A) 111.2
Equals 0.040
Results multiplied by 100 . . . . 0.040 × 100
Equals percent change . . . . 4.0 (% increase above Period A index)

Percent changes for 3-month and 6-month periods are expressed as annual rates (computed according to standard compound growth rate formulas), which mean they indicate what the percent change would be if the current rate were maintained for a 12-month period.

The national CPI–U for March 1992 was 139.3. (Remember, 1982–84 is equal to 100.) Therefore, since 1982–84, there has been a 39.3 percent change—increase—in prices. This means that the typical urban family would have had to spend 39.3% more in March 1992 to maintain its 1982–84 standard of living.

✓*TIP*   Remember that for any particular individual or family, the Consumer Price Index is not necessarily an exact indicator of the effects of price changes. First, the CPI does not count income taxes or Social Security taxes that affect you as a consumer. (The CPI does incorporate all other sales and excise taxes that accompany the market basket items.) Second, the CPI includes items that may *not* affect you (for example, it counts the cost of public transportation in New York City). Finally, integral to the CPI is the idea of the *fixed* market basket (always the same goods in the same quantities), while in reality, individual consumers adjust their purchases to cut down on higher-priced items and take advantage of more moderately priced ones.

Section 2: Word Pictures: Reading Tables and Graphs*

Truly, numbers are inescapable. They are all around us on the pages of daily newspapers and magazines, on radio and television—statistics about crime, inflation, unemployment, industrial productivity, number of TV viewers. Here's a recent sampling:

"NBC's ratings increased 9%"
"Chrysler plans to reduce its worldwide work force by about 14,500 or 21%, by the end of the year"
"NASDAQ issues rose 0.54%, closing at $567.80, up $3.05 for the day."

Examples like these abound. Understanding them requires knowing how percentage increases or decreases are computed. For this we refer you to Chapter 1, Section 1.

Frequently, however, statistical information is not presented in words, but in *word pictures* in the form of tables and graphs. Reading them intelligently does not take sophisticated mathematical knowledge, but it does require a few moments of carefully focused attention. In fact, as we'll show you here, the old saying, "One picture is worth 1,000 words" is really very meaningful when it comes to the presentation and interpretation of data.

Here's how

READING TABLES

---

*The graphs in this section were adapted from *The Math Solution: A Skill Building Course with Business Applications,* by Stanley Kogelman and Victor D'Lugin (1982), AMACOM Special Products Division, New York.

Tables and graphs are pictorial ways of quickly and efficiently presenting large amounts of numerical data in a form that can be readily digested. They are also helpful in demonstrating relationships between two or more sets of facts, and they have dramatic appeal that catches the reader's attention and makes points forcefully.

Suppose, for example, that we wanted to communicate information about the financial status of The Do-Good Company, including information about total annual sales, cost of goods sold, other costs, income before taxes, net income, and amount of dividends declared. Such a wealth of information is hard to absorb when it is presented orally or in writing. Instead, summarizing it, as in the table below, enables it to be read quickly and easily.

Table 1
Summary of Financial Information (1991)
The Do-Good Company
(In millions of dollars)

Annual Sales	$1,304
Cost of Goods Sold	1,069
Other Expenses	111
Income Before Taxes	124
Net Income	96
Amount of Declared Dividends	39

(If you're wondering at this point, "What am I supposed to do with all this information?" the answer is to just read it and absorb what's there.)

To begin to interpret the figures in the table, first notice that it says the numbers are "In millions of dollars." This means that the annual sales figure, for example, which looks like $1,304, is actually $1,304,000,000. That's more than one billion dollars.

Now let's see if the figures check in some way. As you look down the column, you see a breakdown of annual sales. Note that the cost of goods sold (1,069) + other expenses (111) + income before taxes (124) adds up to the total annual sales (1,304). Also, notice that the difference between total annual sales (1,304) and total expenses (1,069 + 111) is equivalent to income before taxes (124).

What else can we tell from these figures? The difference between the income before taxes (124) and the net income (96) is the amount of the taxes (28), which is *not* shown in the table but has to be deducted from the data that is available.

This type of analysis gives you a feel for the figures, but it's not an automatic process. It is the result of careful study of the detailed information that is presented.

Often, information for only a single year doesn't tell very much about a company. In some cases, tables of multi-year summaries provide a better picture of trends—how a company is doing from one year to the next.

Consider the illustrative table below:

Table 2
Five-Year Summary of Financial Information (1987–91)
The Do-Good Company
(In millions of dollars)

	1987	1988	1989	1990	1991
Annual Sales	772	842	860	1,042	1,304
Cost of Goods Sold	556	577	638	806	1,069
Other Costs	90	106	75	118	111
Income Before Taxes	126	159	147	118	124
Net Income	56	62	76	57	96
Amount of Declared Dividend	14	48	79	14	39

It's now possible to follow year-to-year trends in any category listed by reading horizontally across a row from 1987 through 1991. For example, the first row shows what has happened to annual sales over the years. (You could also read down any column to find the complete financial information for any given year. If you find such a vast array of numbers confusing, use a piece of paper to cover up everything except the row or column you are interested in.)

Starting to interpret the data, we find a few items worthy of note:

- Look across the first row and see that annual sales increased year to year.
- Read across the fourth row. Notice that while sales increased from year to year, the income before taxes did not. This is because costs increased faster than sales in some years.
- A glance at the bottom row shows that, in terms of dividends, the best year by far was 1989. This is true despite the fact that annual sales and net income were highest in 1991.

Observations like these are at the heart of reading tables. There's no magic about extracting these facts and trends, just patience with detail.

While tables can present a great amount of information in very compact form, they still require study of the detailed figures in order to ascertain trends. Graphs, on the other hand, let you see trends more quickly and with much less effort.

Here's how

READING GRAPHS

The figure below pictures the 5-year annual sales data of The Do-Good Company in the form of a *line graph*. On this graph, the horizontal axis represents years, and the vertical axis represents millions of dollars in annual sales.

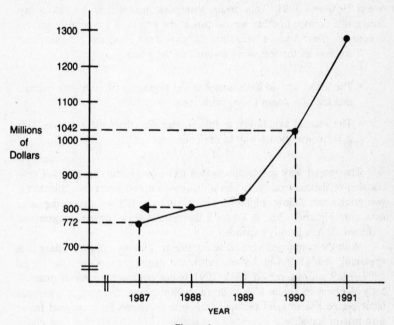

Figure 1
Annual Sales of the Do-good Company (1987-91)

The first point on the line graph stands for 1987 annual sales of $772 (million). The point is located by finding the intersection of a vertical line through 1987 and a horizontal line through the value 772. Similarly, you can plot 1990 annual sales of $1,042 on the graph by finding the point of intersection of a vertical line through 1990 and a horizontal line through 1,042.

It's possible to read information from the graph. The annual sales in 1988, for example, can be found by following an imaginary vertical line from the 1988 mark on the horizontal axis until it hits the line. Then move horizontally until you meet the vertical axis where you'll read sales of about $840. The actual value as we saw from the table was $842, but $840 is certainly a good enough approximation.

While it's more exact to read actual values from a table than from a graph, it's easier to see patterns in the data from a graph. Notice, for example, that the line we have drawn connecting the points in Figure 1 is rising from 1987 to 1991. This means that sales steadily increased during that period. The fastest rise (steepest segment of the line) took place between 1990 and 1991. This means that sales increased at the fastest rate during this period. In other words, this is the period of the greatest percent increase in sales. And we found this all out without doing any computations!

Take note of the following features of the graph:

- The small double hash marks at the beginning of each axis indicate that the axis doesn't start from zero.

- The vertical axis is just a little shorter than the horizontal one. This is the conventional way to draw the axes.

The reason why graphs are drawn this way is that changing the relationship of the axes can give you a distorted impression of the data. In the two graphs that follow, Figures 2 and 3, on page 193 we present the same data as in Figure 1. But, in Figure 2 the vertical scale is greatly contracted; in Figure 3, it is greatly expanded.

With the contracted vertical scale, Figure 2 shows sales still rising, but apparently not as fast as before. Also, the distinction between the period 1987–1989 and the period 1989–1991 is not portrayed nearly as dramatically as it was in Figure 1. In contrast, with the expanded vertical scale, the latter period rise in sales pictured in Figure 3 appears far more impressive than it is in actuality.

✓*TIP*   When reading graphs, take note of the size of the axes. If the vertical axis is about ¾ the length of the horizontal axis, the graph is likely to present a fair picture of the data. If the vertical axis is much longer or shorter than the horizontal axis, be sure to read the numbers on the scales. This will help you to correct the possibly misleading picture presented by the graph.

So far, we've considered data on annual sales for The Do-Good Company in two different forms: a table and a line graph. There is another type of graph that is often used with this type of data. It's called a *bar graph*. The figure below (Figure 4 on page 194) is a bar graph representation of the annual sales of The Do-Good Company. The axes are drawn exactly as in Figure 1 and the points that are plotted on the graph are also located the

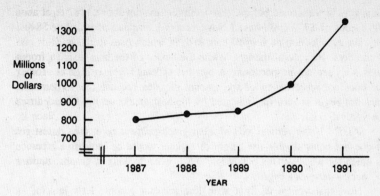

Figure 2
Annual Sales of the Do-good Company (with contracted vertical scale)

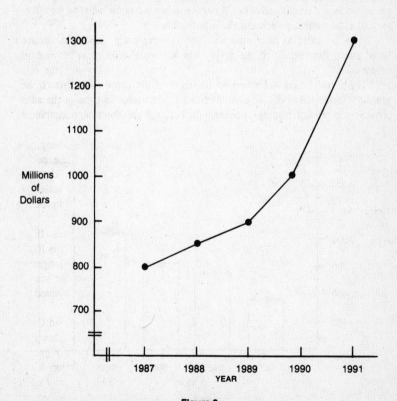

Figure 3
Annual Sales of the Do-good Company (with expanded vertical scale)

same way we described before. (For example, the dot above 1987 represents 1987 sales of $772 (millions).) Now, however, instead of connected dots, we have a collection of vertical bars each of which rises to the level of one of the dots. Be careful though! While the height of the bars does represent sales, they are not proportional. A bar that appears twice as high as another bar does not represent twice the amount of sales because the vertical axis does not begin at zero (as indicated by the hash marks on the lower part of the vertical axis).

✓*TIP*   If the vertical axis of a bar graph starts at zero, then the height of one bar being double the height of another would be indicative of twice as much (of whatever the bars depict.) So, when looking at graphs, *be sure to check where they begin.*

Now that you're familiar with line and bar graphs, let's look at an example of a type of graph commonly found in the newspaper. Reading Figure 5 (see page 195) requires careful study because the axes are not presented in the traditional way. However, since we know what to look for, we can systematically seek out the information.

First, we need to determine what the axes represent. Since months are listed along the bottom of the graph, the horizontal axis must be time in months.

The vertical axis is located on the right of the graph. The title of the graph, "Trading Activity of Lambda Stocks," indicates that the graph must somehow represent trading. Looking further, we see the sentence, "New

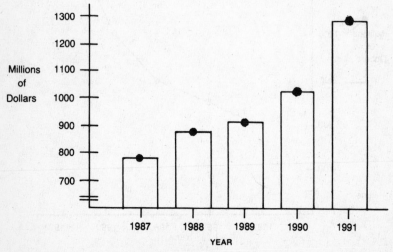

Figure 4
Annual Sales of the Do-good Company

purchase or sales of Lambda stocks listed on all exchanges in millions of dollars.'' This tells us that the right scale measures millions of dollars. The words "Net Purchases" written near one of the peaks shows us that the upper part of the graph (the portion above the zero line) represents purchases. Therefore, the lower part must represent sales.

Now we can see that net purchases were highest in September 1990 (at more than $300 billion). We can also see that there was a shift from net purchases to net sales between September and October. But between October and November, Lambda Stocks again moved to a net purchase situation. This didn't last, however, and the stock dropped to a net sales position for the remainder of the graph (until February 1991).

The last type of graph we'll consider is a *circle graph*. Circle graphs are used to show the breakdown of some fixed quantity into subdivisions. The graph pictured on page 196 represents the average age of students enrolled in advanced mathematics courses in the United States. One hundred percent is being broken down in this instance. One hundred percent refers to 100 percent of the students in these courses.

The circle graph is sliced so that percentages correspond to appropriate portions of the circle. Since 20-year-olds comprise 32.7% of the total, for example, roughly 33% of the circle (that's nearly one-third) is marked off for this age group. Since about 11% of the total is 22-year-olds, this is

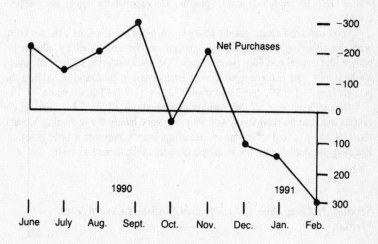

Figure 5
Trading Activity of Lambda Stocks
Net purchase or sales of Lambda stocks listed on all exchanges in milions of dollars.

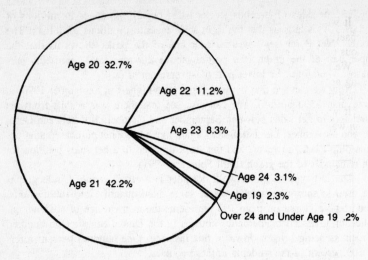

Figure 6
Average Age of U.S. Students Taking Advanced Mathematics Courses

represented by a slice that is one-third as large as the slice apportioned to 20-year-olds.

✓*TIP*   In well-done circle graphs, the size of the slices are proportional.

You can read circle graphs at a glance, slice by slice, or you can combine the slices. For example, age groups can be combined by adding together the corresponding percentages. Of the students taking advanced mathematics, the percent aged 19–21 inclusive can be found by adding the percentages of 19, 20, and 21-year-olds (2.3 + 32.7 + 42.2) to obtain 77.2%.

There is no end to the various combinations of line, bar, and circle graphs that can be produced. And they are very handy for providing a quick view of complicated information. Reading graphs requires a little practice. Reading the really complex ones requires lots of patience as well.

Section 3: Lies, Damned Lies, and Statistics (*Average* vs. *Median*)

Michael's father used to tell him a bedtime story about a ship's captain who took his little son on a sailing trip. To save him from drowning one day when a storm threatened to capsize the boat, the captain strapped him to the mast.

It was not until many years later in a high school English class that Michael learned this story was the plot of "The Wreck of the Hesperus," a famous poem by Henry Wadsworth Longfellow.

We had much the same experience with "averages." We started computing arithmetic averages way back in third or fourth grade. It was one of the first things we did when we got our report card. . . .

It wasn't until much later in college that we learned that average or *mean* was a *statistical concept.*

REPORT CARD	
SUBJECT:	GRADE:
Arithmetic	~~89~~ 95
English	93
History	84
Social Studies	~~95~~ 89
Art	90
French	88
Physical Education	92
	90.14 = AVERAGE

But it was still computed in the same way we did it in fourth grade.
Here's how

AVERAGE or MEAN

We added up all the grades and divided by the number of them we added together.

The mean is defined as the sum of all the scores (or grades or measures) divided by their number. In *statistical notation*, the formula for the mean (which is symbolized by $\overline{X}$ is:

$$\overline{X} = \frac{X_1 + X_2 + \cdots X_n}{n}$$

In this formula:

X_1 stands for the first score (or measure or grade), X_2 stands for the second score $\cdots$ and X_n stands for the "n th" or last score.
n stands for the total number of scores.

When applying this formula to the fourth grade scores, you have $X_1 = 95$, $X_2 = 93$, and so on; and $n = 7$. Then:

$$\overline{X} = \frac{95 + 93 + 84 + 89 + 90 + 88 + 92}{7} = \frac{631}{7}$$
$$= 90.14$$

Notice that the result, 90.14, does not match any of the scores in the distribution of grades. Actually, it's not even a whole number like the other scores. Nonetheless, the mean grade *summarizes* how we performed in the several different subjects. In statistical terms, the mean depicts the central "value" or "tendency" of a distribution or collection of scores in the sense that the scores are *evenly balanced* about the mean. This should not be interpreted to mean that the mean falls in the center of a distribution of scores.

To illustrate how the scores in a distribution balance about the mean, let's compute all the scores' "deviations" from the mean. We do this by subtracting the mean from each score, like this:

SCORE	(MINUS)	MEAN	(EQUALS)	DEVIATION
95	−	90.14	=	4.86
93	−	90.14	=	2.86
84	−	90.14	=	−6.14
89	−	90.14	=	−1.14
90	−	90.14	=	−0.14
88	−	90.14	=	−2.14
92	−	90.14	=	1.86

In this example, note that the negative values ($-$) indicate that the score fell below the mean, while the positive deviations (by tradition, the "$+$" sign is left out) indicate that the score was above the mean.

Now, if we add up the values of the deviations that are below the mean ($6.14 + 1.14 + .14 + 2.14 = 9.56$) and add up the values of the deviations above the mean ($4.86 + 2.86 + 1.86 = 9.58$), we find that they just about balance. In fact, the scores would be *exactly* balanced about the mean if the mean score has not been rounded off to 90.14 (but left as 90.14285714).

The mean is the "balance point" of any distribution of scores. However, it can be very misleading when there are extreme scores. For example, suppose we had gotten 65 in French instead of 88. When we recompute the mean, we find it falls to 86.86 as a result of this one low grade. While the new mean is the balance point of the new distribution in that the sums of the positive and negative deviations balance (or come to zero), you would be mistaken to interpret the mean as the point around which most of the actual scores fall. They do not; actually, all scores fall above this mean except for the just-passing grade we got in French.

Let's consider another example of a distribution with an extreme score. These are the times of six people running a 100-yard race: 9.6, 9.7, 9.8, 9.9, 10.0, and 14.0 seconds. Applying the formula for the mean, we find that:

$$\overline{X} = \frac{9.6 + 9.7 + 9.8 + 9.9 + 10.0 + 14.0}{6}$$
$$= \frac{63.0}{6}$$
$$= 10.5$$

Here you can clearly see that the mean of 10.5 seconds does not satisfactorily describe the typical runner's time because of the distorting effect of the extremely slow racer.

In almost any instance where there are *extreme scores in one direction,* the mean is probably not the statistic that is most "typical" of the distribution. There's another statistic that gives a more typical description in the case of extreme scores.

Here's how

THE MEDIAN

The median is defined as the middle value (score, measure, grade) in a distribution so that half of the scores fall above it and half fall below it. It is also a measure of "central tendency." If there are an *odd* number of values (scores) in the distribution, the median is the value of the middle score. When there is an *even* number of values (scores) in the distribution, the median lies between the two middle scores and is computed to be the average of these two scores.

To calculate the median, you must *first arrange all the scores in order.* Let's stick with the report card, for which we'll arrange the grades from low to high. (We could also arrange them from high to low.)

84   88   89   90   92   93   95

Since there is an odd number in this distribution of seven grades, the middle score is the fourth one. The median is 90. Three scores fall above the median and three scores fall below it.

Let's assume we didn't take French that semester and had only six subjects (an even number), like this:

84   89   90   92   93   95

The median still falls halfway into the distribution, in this case, halfway between 90 and 92. To find the median, add the two middle scores:

$90 + 92 = 182$

And divide by 2:

$182 \div 2 = 91$

The median is 91. The median is thus not one of the actual scores in a distribution with an even number of data points.

Now, we're going to compare means and medians to illustrate when you would want to use one rather than the other.

Here's how

MEAN VS. MEDIAN

Using all seven report card grades, we just calculated the mean as 90.14 and the median as 90. They are very close together, and either of the two statistics satisfactorily describes how we did in school. When we dropped French from the distribution and used only six scores, the mean (90.5) and median (91) were still pretty close together and either one could be used to describe the typical score.

But, in a distribution that has extremely high or extremely low scores, the values of the mean and median are not similar. Consider the times in the 100-yard race:

9.6   9.7   9.8   9.9   10.0   14.0

Here, the mean is 10.5. The median (which falls halfway between 9.8 and 9.9 ([9.8 + 9.9] ÷ 2) is 9.85. In this instance, the median gives a better picture of how most of the racers performed.

Now let's take a look at the salaries of nine people, including the president of the Freehold Company. They earn:

$ 29,700
158,000
46,225
38,750
25,800
35,375
22,175
48,425
50,500

Calculating the mean:

$$X = \frac{454,950}{9}$$
$$= 50,550$$

We find the average salary to be $50,550, which is a higher salary than any of the employees with the exception of the company's president. The mean falls at the very upper end of the distribution, a result we obtained because of the effect of the extreme salary: the president's.

The median, on the other hand, is not so affected. Arranging the scores in order to compute the median (we'll arrange them from high to low as in Case A below), we find the median to be the *fifth value:* $38,750. (This is an odd-number distribution, so the middle score is the median.)

Table 1
Freehold Workers' Earnings

CASE A	CASE B	CASE C
158,000	195,000	75,000
mean (50,550)	mean (54,661)	50,500
50,500	50,500	48,425
48,425	48,425	46,225
46,225	46,225	mean (41,328)
median 38,750	median 38,750	median 38,750
35,375	35,375	35,375
29,700	29,700	29,700
25,800	25,800	25,800
22,175	22,175	22,175

Compare the median of $38,750 with the mean of $50,550. Which statistic shows better what most Freehold Company workers earn?

Because the median is not affected by the size of the extremes, we could raise the president's salary to $195,000 without changing the median

(Case B above.) Similarly, if we reduced the President's salary to $75,000 we would not affect the median (Case C above). *But we would affect the mean in each of these Cases.* When we increased the amount the president earned, we increased the mean salary to $54,661. When we reduced the president's salary so that it wasn't so much higher than the other salaries, the median was not changed, but the mean dropped to $41,328. In the case of the lowered president's salary, the effect of reducing the extreme score was to bring the mean and median into alignment.

But it's not only the mean that can give a distorted view. The median can too. For example, suppose you were dealing with the following salaries:

$ 15,000
20,000
25,000
mean (57,500)
60,000
median (65,000)
70,000
80,000
90,000
100,000

In this distribution, the median is 65,000 ([60,000 + 70,000] ÷ 2), while the mean is 57,500. Neither statistic gives a good picture of the salary structure of this company—that is, of the wide range of salaries—but consideration of *both* the mean and the median makes it more apparent that there are many very high salaries, but also some low ones. Here's why. To better understand this, think about how you could have a median of $65,000 and a mean of $57,500. A median of $65,000 implies that half of the scores were $65,000 or more (and half were less). But the mean was quite a bit lower than this amount. Therefore, there must have been some very low salaries to pull the mean down that low.

✓*TIP*   With salaries in particular, where there tend to be extreme scores in one direction, the median is probably a better indicator of typical earnings. In fact, when you see the mean (or average) salary of a company or department reported, you might well question whether there is some distortion resulting from the use of this statistic. After all, if we wanted to make the Freehold Company look like it pays *all* its employees high salaries, we would report the *mean* salary. These are the kind of "lies" Benjamin Disraeli meant when he talked about ". . . lies, damned lies and statistics."

10

Kitchen
Math

Section 1: Timely Thoughts, or "We're Having Company for Dinner"

"Let's have Thanksgiving dinner at our house!"

What, you ask, does cooking a holiday dinner for 18 people have to do with math?

Plenty!

First, we need to figure out how large a turkey to buy. Then, we need to calculate how long to cook the bird. And, finally, we have to know whether to double or triple our stuffing recipe to make sure we have enough.

Here's how

BUYING A BIRD: POUNDS/PERSON

A good rule of thumb is to allow from three-quarters to one pound of turkey per person, because of the bones. When you're buying steak (or fish where the bones are negligible), the per person allowance is one-half pound. So, for 18 guests, at ¾ pound per person, we should order a 13 to 14-pound turkey. To arrive at the total weight, multiply the per pound allowance by the number of people:

$$\tfrac{3}{4} \times 18 = 0.75 \times 18 = 13.5 \text{ lbs.} \quad \left\{ \begin{array}{l} \text{Change the fraction } (\tfrac{3}{4}) \text{ to a decimal} \\ \text{by dividing the numerator } (3) \text{ by the} \\ \text{denominator } (4). \end{array} \right.$$

If we allow one pound per person, we would buy an 18-pound turkey.

To be on the safe side, and to make sure we have leftovers for our family and enough for doggy bags for guests, we'll order a 20-pound bird. This should feed 20 very hungry people (at one pound per person) or 26–27 moderately hungry guests at ¾ pound person, like this:

$$20 \text{ lbs.} \div \tfrac{3}{4} \text{ lb/person} = 20 \div 0.75 = 26.7 \text{ people}$$

Twenty pounds will certainly be more than adequate for 18 holiday dinners, plus next day turkey soup, sandwiches, and turkey burgers.

With fresh turkey selling at $1.29 per pound, what is the cost of a 20-pound turkey?

$$\$1.29 \times 20 = \$25.80$$

To carry this one step further, we can compute the per person cost by dividing the total cost by the number of *diners:*

$$\$25.80 \div 18 = \$1.43 \text{ per person}$$

Or we can compute the per serving cost by dividing the total cost by the expected number of *dinners:*

$$\$25.80 \div (18 + \text{leftovers for } 6) = 25.80 \div 24 = \$1.08 \text{ per serving}$$

Turkey is really quite an inexpensive dish. How does it compare with steak, for example, which is selling for $3.59 per pound? At one-half pound per person, to feed 18 people, we would need 9 pounds of steak. Nine pounds of steak will cost $32.31 ($9 \times \$3.59$), which represents a per person cost of $1.80 ($32.31 \div 18$).

Steak compares so favorably to turkey when calculated on a per person basis because of the difference in the amounts suggested per person and because so much "extra" is generally not built in for leftovers.

The next question is, now that we have it home, for how long must we roast the turkey? Our cookbook indicates 20 minutes per pound in a moderate (325° to 350°) oven.

Here's how

COOKING TIME

To compute the total cooking time, multiply the recommended minutes per pound by the total number of pounds:

20 lbs × 20 min/lb = 400 minutes

Now, divide the total minutes by 60 (there are 60 minutes per hour) to get the total cooking time in hours and minutes:

400 ÷ 60 = 6.66

This is *not* 6 hours and 66 minutes but, rather, 6 hours and 0.66 of an hour.

✓*TIP* To compute a fraction of an hour, multiply the fraction (or decimal) by 60 minutes:

0.66 × 60 = 39.6 or, rounding off, 40 minutes

So, 400 minutes equals 6 hours and 40 minutes. We can check this by multiplying 6 hours by 60 minutes (6 × 60 = 360) and then adding the 40 minutes (360 + 40 = 400).

Let's do some other examples.

EXAMPLE: How long will it take to roast an 18-pound turkey?

18 × 20 = 360 minutes
360 ÷ 60 = 6 hours of cooking time
    (check: 6 × 60 = 360)

EXAMPLE: A 12-pound turkey will require 4 hours of cooking time at 20 minutes per pound:

12 × 20 = 240
240 ÷ 60 = 4 hours

EXAMPLE: And a 13½-pound turkey should cook for 270 minutes. This is 4½ hours:

13.5 × 20 = 270
270 ÷ 60 = 4.5, which is 4 hours and one-half hours.

Going back to our original example (the big bird), we need 6 hours and 40 minutes of cooking time. Assuming we want to serve dinner at about 4:00 in the afternoon, at what time do we put the turkey in the (pre-heated) oven?

There are basically two methods for subtracting hours and minutes (and for adding time) when both times are in the AM or both are in the PM.

Here's how

SUBTRACTING TIME

Let's start with *Method 1*.

Step 1 Set up the problem as a subtraction problem with hours and minutes listed side by side. For example, the time difference between 8:40 AM and 11:20 AM is set up as follows:

$$
\begin{array}{rll}
& 11 \text{ hours} & 20 \text{ minutes} \\
- & 8 \text{ hours} & 40 \text{ minutes} \\
\hline
\end{array}
$$

Step 2 In this example, since 40 is "bigger" than 20, it is necessary to borrow 1 hour (60 minutes) from the 11 hours and add it to the 20 minutes:

$$
\begin{array}{rllllll}
& 11 \text{ hours} & 20 \text{ minutes} & = & 10 \text{ hours} & 80 \text{ minutes} \\
- & 8 \text{ hours} & 40 \text{ minutes} & = & 8 \text{ hours} & 40 \text{ minutes} \\
\hline
\end{array}
$$

Step 3 Do an ordinary subtraction:

$$
\begin{array}{rll}
& 10 \text{ hours} & 80 \text{ minutes} \\
- & 8 \text{ hours} & 40 \text{ minutes} \\
\hline
& 2 \text{ hours} & 40 \text{ minutes} \\
\end{array}
$$

The result is the time difference.

Method 2 is called *Use Your Fingers!*

✓✓*HOT TIP*   Yes, counting on your fingers is highly recommended. For many calculations, it is the preferred procedure.

Step 1 Start by counting the number of minutes from the earlier time to the *next full hour*. (For example, from 8:40 to 9:00 is 20 minutes.)

Step 2 Count the number of hours from this next full hour to the *last full hour*. (For example, from 9:00 to 11:00 is 2 hours.)

Step 3 Count the number of minutes from the last full hour to the *final time*. (For example, from 11:00 to 11:20 is 20 minutes.)

Step 4 Add together the results from Steps 1, 2, and 3 to find the total time difference. (For example, 20 minutes + 2 hours + 20 minutes = 2 hours 40 minutes.)

Our real-life cooking problem is a little different. Here, we know how long it takes to roast our turkey, (6 hours and 40 minutes), and we know the time we want it to be finished (by 4:00 PM). What we are looking for is the *starting time*. We need to count back 6 hours and 40 minutes from 4:00 PM.

Let's first use *Method 2* and count backwards on our fingers.

From 4:00 PM backwards to noon equals 4 hours. Counting back 2 more hours, from noon to 10:00 AM, gives us a total of 6 hours. We need another 40 minutes:

$$\begin{array}{r} \underline{-\ \ 10 \text{ hours and }\ 0 \text{ minutes}} \\ 40 \text{ minutes} \end{array} = \begin{array}{r} \underline{-\ \ 9 \text{ hours and } 60 \text{ minutes}} \\ 40 \text{ minutes} \\ \hline 9 \text{ hours and } 20 \text{ minutes} \end{array}$$

We have to put the turkey in the oven at 9:20 AM.

✓*TIP* We recommend that you actually put the turkey in 15 to 20 minutes earlier, so at the end, when the bird is fully cooked, you can let it "rest" outside the oven. This 15 to 20 minute rest period allows the juices to set and makes it easier to carve the turkey.

Here's how we would do it with a 24-hour clock.

24-HOUR CLOCKS

A 24-hour clock is not so much a physical apparatus as another way of counting the hours from midnight to midnight without using AM and PM. Both the 24-hour clock (the clock used by the military because it offers less chance for confusion) and the 12-hour clock (the clock we normally use) start at midnight. Therefore, the morning times are the same on both clocks: 10:15 AM or 1015 means 10 hours and 15 minutes past midnight; 11:30 AM (1130) means 11 hours and 30 minutes past midnight; and 12:00 noon (1200 hours) means 12 hours past midnight.

The two types of clocks differ in how they present the times *after* noon. For example, 1:30 PM (which on a regular clock is one and one-half hours after noon) is 13 hours and 30 minutes past *midnight* on a 24-hour clock.

This would be presented as 1330. As another illustration, 3:00 PM is equal to 15 hours after midnight, or 1500 on a 24-hour clock. Thus, *to convert a* PM *time to a 24-hour clock time, just add 12.* Here are some additional examples:

  7:35 PM = 1935 hours
10:42 PM = 2242 hours
11:59 PM = 2359 hours

To convert times greater than 1200 on a 24-hour clock to a 12-hour clock, just subtract 1200 to arrive at a PM reading. Going back to our cooking problem, if we wanted the turkey ready at 1600 hours (4:00 PM), and it had to cook for 6 hours and 40 minutes, by *Method 1,* subtraction, we'd have to put it in the oven at 0920 (9:20 AM):

$$\begin{array}{r} 16 \text{ hours and } 0 \text{ minutes} \\ - \ \ 6 \text{ hours and } 40 \text{ minutes} \\ \hline \end{array} = \begin{array}{r} 15 \text{ hours and } 60 \text{ minutes} \\ - \ \ 6 \text{ hours and } 40 \text{ minutes} \\ \hline 9 \text{ hours and } 20 \text{ minutes} \end{array}$$

We use more math when we prepare the stuffing:
Here's how

INCREASING RECIPES

The New York Times Cookbook (Craig Claiborne, ed., New York: Harper & Row, 1961) indicates that about ¾ to 1 cup of stuffing is needed for each pound of ready-to-cook bird. It then proceeds to present a "Basic Bread Crumb Stuffing" recipe for a 5-pound bird, as follows:

One small onion
One stalk of celery with leaves, chopped
⅓ to ½ cup butter (or margarine)
1 to 2 teaspoons poultry seasoning or sage
Freshly ground black pepper

2 tablespoons chopped parsley (optional)
5 cups of stale bread cubes or crumbs
½ teaspoon salt
Water, milk, or giblet gravey (optional)

We need to quadruple (multiply to 4) these ingredients to fill our 20-pound turkey. (Here, it might be helpful to refer to Section 3 of this chapter.)

one small onion × 4 = 4 small onions or 2 medium-sized ones
one stalk celery × 4 = 4 stalks of celery
(⅓ to ½) cup butter × 4 = 1⅓ to 2 cups butter
1 to 2 teaspoons poultry seasoning × 4 = 4 to 8 teaspoons *
freshly ground black pepper × 4 = only taste will tell! *
2 tablespoons of chopped parsley × 4 = 8 tablespoons or ½ cup
5 cups of bread cubes × 4 = 20 cups
½ teaspoon salt × 4 = salt to taste! *
liquid × 4 = enough to barely moisten bread

To finish the recipe:

1. Sauté the onion and celery in the butter or margarine until tender but not brown.
2. Combine the seasonings and bread crumbs, toss together with the onion mixture and, if a moist dressing is desired, add enough liquid to barely moisten crumbs.

Then, lightly fill the cleaned and salted body cavity and the neck or wishbone cavity: do not pack the dressing since it greatly expands in cooking. Close the cavities (with skewers and/or by sewing); place the bird on its back in a pan; cover the breast with generous amounts of butter or margarine or with a cloth soaked in melted fat; pop it into a moderate oven; and cook for 6 hours and 40 minutes, basting often with the drippings from the pan.

Dinner will be ready at 1600 hours.

Section 2: Not So Small Differences: Comparison Shopping

We just returned from a grocery shopping expedition that confirmed our suspicions that it would be difficult to decide upon the best buy. For example, we found three bags of the same brand, same type of potato chips priced as follows:

WEIGHT	PRICE
7½ oz.	$1.39
11 oz.	$1.99
16 oz.	$2.49

---

*Most spices, but especially salt, are not increased proportionally. Use judgment and taste to figure out how much more you need.

Odd quantities such as 7½ ounces or 11 ounces make it hard to compare prices in your head. If, for example, the quantities had been 8, 12, and 16 ounces, the comparisons would be quite straightforward. You would expect the price of the 16-ounce bag to be twice the price of the 8-ounce bag since it is twice the weight. If the price was less than double, the larger package would be a good buy.

While it's not as easy to compare a 12-ounce bag with an 8-ounce bag as it is to compare 16 ounces and 8 ounces, it's still not too bad. You could reason this way: a 12-ounce bag is one and one-half times the size of an 8-ounce bag, so the price of a 12-ounce bag should be one and one-half times the price of an 8-ounce bag. If the price is lower than this, the 12-ounce package is the better buy; if it's higher, buy the smaller bag of chips.

In the market we visited, odd quantities and uneven prices, such as those on the potato chip packages, were the rule rather than the exception. The only reasonable way to comparison shop given such incompatible weights and such "hard" pricing structures is with a calculator. That's the way we recommend you determine *unit prices*, although, at first thought, bringing a calculator to the supermarket seems terribly embarrassing. In fact, current pricing is designed to discourage you from comparison shopping in a rigorous way. So, do consider using a calculator. It takes only a moment to compare prices with it. . . . And it can really save you money.

Here's how

UNIT PRICING

Prices of different amounts of similar items can be compared by computing the price per pound, per ounce, per quart, or any other measurement unit. To figure out the cost per unit of weight (or volume), divide the price of the item by the quantity that you can buy for that price. Expressed as a formula:

$$\text{Unit price} = \frac{\text{Price}}{\text{Quantity}}$$

The required division is best done on a calculator.

✓*HOT TIP*   The goal of comparison shopping is to determine the best buy. *This is the item with the lowest unit price.* However, you may sometimes decide to buy a smaller or a larger size box (or bag or bottle) than the one with the lowest unit price because of other factors, such as spoilage or convenience. As an illustration, a gallon of milk may be a better buy than a quart of milk (it also may not be), but if you use only a little milk, the gallon size might spoil before you can come close to finishing it. Similarly, the limitations of home storage space may also mitigate against purchasing giant-sized packages.

Now let's look at some examples. These are not "textbook" cases but are the actual prices we found in a large local supermarket on a Saturday morning, not too long ago.

EXAMPLE: Let's start by figuring out which bag of potato chips is the best buy:

7½ oz. for $1.39
11 oz. for $1.99
16 oz. for $2.49

Solving this problem means that we have to find the unit price for each of the three bags. To repeat the *unit price formula:*

$$\text{Unit price} = \frac{\text{Price}}{\text{Quantity}}$$

Substituting in the formula, the unit price for the smallest bag is:

$$\text{Unit price} = \frac{\$1.39}{7.5 \text{ oz.}}$$
$$= \$0.185 \text{ per oz.}$$

Translating this into cents by multiplying by 100 (100 cents in a dollar), the first bag costs 18.5¢ per oz.

By calculator:
PRESS 1.39 ÷ 7.5 × 100 =

For the middle-size bag, the unit price is:

$$\text{Unit price} = \frac{\$1.99}{11 \text{ oz.}}$$
$$= \$0.181 \text{ per oz.}$$
$$= 18.1¢ \text{ per oz.}$$
By calculator:
PRESS 1.99 ÷ 11 × 100 =

And the largest bag's unit price is:

$$\text{Unit price} = \frac{\$2.49}{16 \text{ oz.}}$$
$$= \$0.156 \text{ per oz.}$$
$$= 15.6 \text{¢ per oz.}$$

By calculator:
PRESS 2.49 ÷ 16 × 100 =

The largest bag has the lowest unit price by about 2½ cents per ounce.

✓*TIP*   Unit price labels appear on market shelves by law, but anyone who shops regularly knows that they are hard to read, hard to find, or missing altogether.

Is the best buy the largest package of potato chips? Yes, in the sense that it's cheapest, ounce for ounce. (There really is very little difference between the unit cost of the 7½-ounce bag and the 11-ounce bag (0.4 cents per ounce). But unless you expect to use a large quantity of chips, you should probably consider one of the smaller sizes. Potato chips get soggy quickly and, besides, having a very large bag around might just be too much temptation.

Intuitively, or because you've often been told so, you may have felt that the biggest size would be the best buy. While it turned out to be the case with Saturday's pricing of potato chips at our market, it's not always the case, as we will now see.

The same Saturday in the supermarket, paper towels were on sale in two different packages. You could get one jumbo role for 93¢ or a package of two smaller-looking rolls for $1.25. A careful reading of the labels revealed that the jumbo roll contained 90 sheets of two-ply towels, which measure 11 inches by 10.6 inches, for a total area of 73 square feet. The two-roll package contained a total of 100 square feet, including 62 two-ply towels per roll. Each sheet in this package also measured 11 inches by 10.6 inches. Which package was the better buy?

✓*TIP*   *All* paper towels rolls are the same width in order to fit the dispensers. The length of a sheet, as well as the number of sheets in a roll, may differ.

Given so much information it's difficult to sort out the relevant from the irrevelant to find a common basis for comparison. Best to use the total square footage because that is a true measure of what's in the package. (But you could also use the number of sheets in the packages, if the sheet size is the same, since both packages give this information and, in this example, both contain two-ply sheets of the same dimensions.)

Since the two-roll package contains 100 square feet and costs $1.25, the unit price can be computed as follows:

$$\text{Unit price} = \frac{\$1.25}{100 \text{ sq. ft.}}$$
$$= \$0.0125 \text{ per sq. ft.}$$
$$= 1.25\text{¢ per sq. ft.}$$

For the jumbo roll, we have:

$$\text{Unit price} = \frac{93\text{¢}}{73 \text{ sq. ft.}}$$
$$= 1.27\text{¢ per sq. ft.}$$

The 0.02 cents per square foot difference in the unit prices is really insignificant, so it makes no difference (price wise) which pack you purchase.

Let's do a third example.

EXAMPLE: We looked at mayonnaise, which was available in three sizes:

32 oz. for $1.99
48 oz. for $2.99
One gallon for $7.99

Which size is the best buy?

Notice that these sizes lend themselves to very easy comparison. Remember that 32 ounces equals one quart, so one quart is about $2. (Note that we have rounded off.) The next larger size, 48 ounces, is 1½ quarts $(48 \div 32 = 1\frac{1}{2})$, so it should cost about one and one-half times as much as the smallest size. The middle size is just about $3, which is one and one-half times $2. So the small and medium-sized jars really cost the same. You might think that the gallon size would be a great bargain (assuming, of course, that you could use so much mayo), but it's not.

One gallon is the same as 4 quarts. The very large size costs exactly 4 times the one quart size $(4 \times \$2)$. Here again, the largest size was no bargain. The main reason to buy this size is convenience, or because you use vast quantities (maybe you have a very large family or a restaurant) and don't want to keep several smaller jars in stock, or maybe you don't want to shop for mayonnaise often.

The examples we just considered were chosen more or less randomly from the supermarket shelves. The same principles apply whether you're buying corn flakes or cookies, butter or bug spray. And while convenience, brand loyalty, or color preference are all legitimate approaches to decision-making, it really pays to first compute unit prices!

Section 3: Equalities: Measure for Measure

When you are in the middle of cooking some complicated dish, it's annoying to find that the recipe either calls for a measurement you don't have readily accessible (for example, *our* measuring cup is marked only in quarters and thirds, not in eighths) or for an amount that you are not generally familiar with. Then, in the midst of all the preparations, you have to stop and figure out a way to decode what a jigger is or to measure out seventh-eighths of a cup.

What we're presenting first in this section is a listing of common measurements in more than one form. Some of this information may be found at the back of your all-purpose cookbook; in developing the guides below, we used several different cookbooks to compile the information.

✓*TIP*   Whenever you see a measurement, keep in mind that, unless it specifically calls for a "heaping" teaspoon, all measurements are given for level amounts.

Let's start with what equals what, beginning with small quantities and moving up to larger amounts.

Here's how

TABLE OF WEIGHTS AND MEASURES

✓*TIP*   When a recipe calls for a "few grains," it means less than ⅛ of a teaspoon; this is equivalent to a "pinch." Here are some more household measurements and their equivalents:

HOUSEHOLD MEASURE	EQUIVALENT HOUSEHOLD MEASURE	APPROXIMATE EQUIVALENT WEIGHT OF WATER IN GRAMS	APPROXIMATE EQUIVALENT WEIGHT OF WATER IN OUNCES (SOLID MEASURE)
80 drops	one teaspoon (tsp)	5 grams	0.2
one teaspoon	⅓ tablespoon	5 grams	0.2
one tablespoon (tbl)	3 teaspoons	15 grams	0.5
2 tablespoons	one (fluid) ounce	30 grams	1.1
one jigger (3 tablespoons)	1½ fluid ounces	45 grams	1.6
¼ cup	4 tablespoons	60 grams	2.1
⅓ cup	5 tablespoons + 1 teaspoon	80 grams	2.8
⅜ cup	6 tablespoons	90 grams	3.1
½ cup (1 gill)	8 tablespoons	120 grams	4.2
⅝ cup	½ cup + 2 tablespoons	150 grams	5.3

HOUSEHOLD MEASURE	EQUIVALENT HOUSEHOLD MEASURE	APPROXIMATE EQUIVALENT WEIGHT OF WATER IN GRAMS	APPROXIMATE EQUIVALENT WEIGHT OF WATER IN OUNCES (SOLID MEASURE)
⅞ cup	¾ cup + 2 tablespoons	210 grams	7.3
1 cup (½ pint)	16 tablespoons	240 grams	8.4 (½ lb.)
2 cups	one pint (16 fluid ounces)	480 grams	16.7 (1 lb.)
one quart	2 pints (4 cups)	946 grams	33.4 (2 lb.)
one gallon	4 quarts (16 cups)	3,784 grams	133.5 (8 lb.)
one peck (Dry measure)	8 quarts (Dry measure)		
one bushel (Dry measure)	4 pecks (Dry measure)		

In addition to tables of weights and measures, other interesting things that you will often find tucked away in the back of cookbooks may include "household hints," ways to clean common stains, "serving suggestions," table service and menu planning, and "definitions" of cooking terms. The tables of equivalents below show the weights of selected foods.

Here's how

TABLE OF EQUIVALENTS

Did you ever wonder how much a stick of butter weighed? One stick weighs 4 ounces or one-quarter of a pound. It is also the equivalent of ½ cup, or 8 tablespoons. Different foods have different weights, as you can see by glancing at the following table.

Table 1
Household Measures and Equivalent Weights

HOUSEHOLD MEASURE	EQUIVALENT WEIGHT
one cup of dried beans	½ pound
one cup of chopped nuts	⅘ pound, shelled
2 cups of butter	one pound
2 cups of cottage cheese	one pound
2 cups of granulated sugar	one pound
2 cups of water	one pound
2½ cups of shortening	one pound
2½ cups of uncooked rice	one pound
3 cups of dried apricots	one pound
3 cups of uncooked macaroni	one pound
4 cups of cocoa	one pound
4½ cups of sifted cake flour	one pound
5 cups of grated cheese	one pound

Whether food is cooked or raw also affects the quantities involved, as illustrated below:

one cup of raw macaroni	= 2 cups of cooked macaroni (the rule is that macaroni doubles, although some cookbooks say pasta grows by ⅓)
one cup of raw rice	= 3 to 4 cups of cooked rice
one cup of whipping (heavy) cream	= 2½ cups of whipped cream

Other equivalencies that you might find useful are:

3 small eggs	= 2 large eggs
5 whole eggs	= one cup
8 egg whites	= approximately one cup
16 egg yolks	= approximately one cup
one square of cooking chocolate	= one ounce
juice of one lemon	= approximately 2 to 3 tablespoons
juice of one orange	= approximately 6 to 8 tablespoons
one grated orange rind	= one tablespoon

Many tables of equivalents and lists of "substitutions" (such as using honey instead of molasses, or baking soda and cream of tartar for baking powder) date back to the time, not too long ago, when it was not uncommon for women to bake cakes and breads as a matter of routine. Then, especially, it was crucial to be able to determine which items could be substituted for others and what were the weights and measure of the ingredients.

Today, however, there is considerably less home baking and the newer, "gourmet" and speciality cookbooks often omit both lists of "equivalents" and "substitutions." Instead, you may find such things as conversion tables for foreign equivalents (see Chapter 6, Section 3—Metrics) and approximate can sizes.

Here's how

CAN SIZES

Obviously, different sized cans weigh varying amounts and contain different quantities. Did you know that there was a time when cans were stan-

Table 2
Standard Can Sizes and Equivalents

CAN SIZE	WEIGHT	CONTENTS
6 ounces (frozen juice)	6 ounces (171 grams)	about ⅔ cup
6½ ounces (tunafish)	6½ ounces (184 grams)	about ¾ cup
No. 1	11 ounces (312 grams)	1⅓ cups
12-ounce can	12 ounces (340 grams)	1½ cups
No. 303	16 ounces (454 grams)	2 cups
No. 2	20 ounces (567 grams)	2½ cups
No. 2½	28 ounces (794 grams)	3½ cups
No. 3	33 ounces (936 grams)	4 cups
No. 10	106 ounces (3,005 grams)	13 cups

dard sizes and also numbered? Presented above is a list of what used to be considered standard sized cans and their contents.

Numbered cans with nice, neat, even amounts of food (like 12 or 16 ounces) are rapidly disappearing. In their place are slightly smaller cans containing odd-sized amounts and their metric equivalents; 19 ounces (1 pound 3 ounces or 539 grams) instead of a No. 2 can; 27 ounces (one pound 11 ounces or 765 grams) replacing the No. 2½ can; and 13.5 ounces (383 grams) that appear to approximate the size of a one pound No. 303 can.

We could find no *good* explanation for the slightly diminishing size of canned goods; the real reason seems to be that this is a way to increase prices without an obvious price increase: selling a 13.5-ounce can for the same price as a 16-ounce can is an example of this kind of marketing strategy. Take a look at some of the cans on your shelf and notice that not only are they smaller than the once standard cans, but they also contain *odd* amounts. Packaging goods in 19 ounce amounts, or 13.5 ounces, for example, certainly makes finding the unit cost of the item that much more difficult, as we saw in Section 2.

11

Home
Improvement

Section 1: Let's Cut a Rug: Computing Area

Did you ever read the label on a can of paint? Typically, it will say something like, "spreading rate is 400–450 square feet per gallon." Area is always measured in square units, such as square inches, square feet, square yards, and so on—except for acres, which are already squared units of measure. Buying carpeting, paint, or wallpaper requires an understanding of how *area* is measured.

Before we get to actually measuring rooms, let's review the concept of area. Then we will show you how to do the computations you need to buy the materials to decorate.

The area of a flat region such as a wall or floor *is the number of square units that it would take to fill that region.* For example, the drawing below represents the floor of a rectangular room that is only 6 feet wide and 8 feet long. The square drawn next to the room measures one foot on a side and is called a *square foot*. The area of the room is the number of square feet you would need to use to cover the floor.

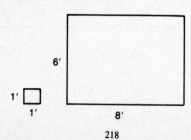

Here's how:

FINDING AREA OF RECTANGLES

To find the area of a rectangle like the one pictured above, *multiply the length by the width.*

> EXAMPLE: Find the area of a rectangle that is 6 feet wide and 8 feet long.
> SOLUTION: Area = Length × Width
> = 8 ft. × 6 ft.
> = 48 sq. ft.

The diagram below illustrates how 48 square feet fit inside the rectangle.

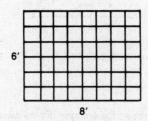

The area of a square is computed in a similar manner.
Here's how

FINDING AREA OF SQUARES

To find the area of a square, multiply the length of a side by itself.
Notice that finding the area of a square is really the same as finding the area of a rectangle because a square is just a rectangle with all sides equal. There also are ways to find the area of other-shaped figures.
Here's how

FINDING AREA OF TRIANGLES

To find the area of a triangle, take one-half the product of the base times the height.

This can be written as the formula:

$$A = \frac{1}{2} \times B \times H$$

A stands for area, B stands for the base, and H is the height.

The height of a triangle is the distance from a vertex (corner) to the side opposite that vertex. *That* side is called the base. It does not matter what pair of sides and vertex you use—the area is always the same.

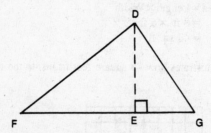

In this diagram, D is one vertex (F and G are the others). The height of the triangle is represented by the dotted line DE. To measure the area of this triangle using the D vertex, use the base side FG.

EXAMPLE: Find the area of the triangle illustrated below.

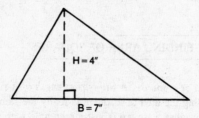

SOLUTION: In this triangle, the height is 4 inches and the base is 7 inches. Therefore:

$$A = \frac{1}{2} \times B \times H$$
$$= \frac{1}{2} \times 7 \text{ in.} \times 4 \text{ in.}$$

$$= \frac{1}{2} \times 28 \text{ sq. in.}$$
$$= 14 \text{ sq. in.}$$

Now we'll show you how to find the area of a circle.
Here's how

FINDING AREA OF A CIRCLE

Finding the area of a circle requires using the following formula:

$$\text{Area} = \pi \times R^2$$

In this formula, R is the radius of the circle, and the Greek letter π (pi) stands for the number 3.14 (approximately). Taken to nine decimal places, π is:

$$= 3.141592654$$

(Also, R^2 is read "R squared" and means $R \times R$.)

EXAMPLE: What is the area of a circle that has a radius of 5 inches?

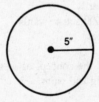

To solve this problem, substitute in the formula:

$$\text{Area} = \pi \times R^2$$
$$= 3.14 \times 5 \text{ in.} \times 5 \text{ in.}$$
$$= 78.5 \text{ sq. in.}$$

The area of a circle with a 5-inch radius is 78.5 square inches.

Sometimes it's necessary to convert from one measurement of area to another. For example, carpeting is generally priced by the square yard. If you've computed the area in square feet, it will be necessary to convert them to square yards.

Here's how

CONVERTING AREA MEASURES

These are the rules for the most common conversions:

A. *To convert square feet to square yards, divide the number of square feet by 9.* There are 9 square feet in a square yard because a square yard measures 3 feet by 3 feet. This yields an area of 9 square feet.

B. *To convert square yards to square feet, multiply the number of square yards by 9.* (This computation is the inverse of the one described in (A) above.)

C. *To convert square inches to square feet, divide the number of square inches by 144.* There are 144 square inches in a square foot because a square foot measures 12 inches by 12 inches; 12 inches multiplied by 12 inches equals an area of 144 square inches.

D. *To convert square feet to square inches, multiply the number of square feet by 144.* (This computation is the inverse of the computation described in (C) above.)

✓*TIP*   By the way, we just thought you might like to know:

one acre = 4,840 square yards
one square mile = 3,097,600 square yards
one square mile = 640 acres

Now that we've reviewed the concept of area, let's turn to some practical applications, like carpeting and paint.

Here's how

MEASURING FLOOR AREA

Even if you're not going to lay carpet or tile yourself, it's a good idea to have some notion of how much carpet (or tile) will be needed for the job. Let's start by assuming that the room you are going to carpet is rectangular.

First, measure two adjacent sides of the room, *including* in your measurement all open spaces such as doorways and closets. Suppose one side measures 11 feet 9 inches and the adjacent side measure 9 feet 8 inches.

Next, find out from the store or manufacturer the width of the carpeting you are interested in. Most carpeting, for example, comes in 12-foot widths although wider widths are available in very expensive carpeting and by special order.

✓*TIP*   If the direction that the pattern will go when laid on the floor does not matter, consider as the *width* of the room the side that is closest to but less than the width in which the carpet is sold. The other dimension (that is, the other side of the room) is the *length* of carpeting you will need to order.

The measurements given in the example above, 11 feet 9 inches by 9 feet 8 inches, indicate that the longer dimension of the room is close to the 12-foot width in which the carpet is sold. Therefore, only a 9 foot 8 inch length of carpet is needed to cover the room. Don't cut it too close! Allow some extra. You probably wouldn't be able to order "and 8 inches" anyway, so order 10 feet.

✓✓*HOT TIP*   If the width in which the carpet is sold is smaller than either dimension of the room, of if the pattern is asymetrical and direction is important, then it will be necessary to buy two runs of carpet, have them sewn together, and then trimmed to the proper width.

To compute the amount of carpeting to be ordered, find the area by multiplying the length and width of the *carpeting*. In the example we are using, the length is 10 feet and the width is 12 feet, so the area of the carpeting we would need is 120 square feet. (Compare this to the area of the *floor*, which measures 11 feet 9 inches or 141 inches by 9 feet 8 inches or 116 inches; 141 inches by 116 inches equals 16,356 square inches, divided by 144 equals 113.6 square feet.)

Since carpet is sold in square *yards*, it is now necessary to convert the number of square feet to square yards by dividing by 9. In our example, it is $120 \div 9$ equals 13.3 square yards. This is the amount of carpeting for which you will have to pay.

Measuring for wallpaper is similar in theory to measuring for carpeting, but in actuality it's quite a bit more complicated. You need to know the width and height of your walls and the width and lengths of the rolls in which paper is sold. You also need to figure in the size of the "repeat"— the amount of space one whole pattern takes up. Since patterns have to be carefully matched, you may have to buy much more paper than you might think. Unless you are very handy, we suggest you leave wallpaper measurement (and hanging!) to the paperhanger or to the wallpaper specialty store manager. Usually as part of their service, they will come to your house to measure.

Paint will cover the area of your walls or ceiling at a predetermined rate that is generally indicated on the paint can. A typical spread rate is

400–450 square feet per gallon. This means that one gallon of paint will cover an area of from 400 to 450 square feet. How much paint will you need to buy for the bedroom?

Here's how

MEASURING FOR PAINT

To determine how much paint you will need:

(1) *Measure the length, width, and height of the room.* Round each measurement up to the next largest footage. Be generous. Don't subtract out windows, doors, or closet space.
(2) The next step involves *computing the area of each wall* by multiplying its width (in feet) by the height of the room (also in feet).
(3) *Add together the areas of the walls* to get the total wall area.
(4) To figure out how much paint to buy for one coat, *divide the total wall area to be painted by the spread rate of the particular type and brand of paint you are planning to use.*

The answer is the number of gallons of paint you'll need for one coat on the walls. Remember, don't cut it too close. Always err on the high side. It's better to have some paint left over (for touch-ups) than to run out in the middle of the job. If you want to apply *two* coats of paint, remember to double the number of gallons of paint you need to buy.

If you are painting the ceiling the same color as the walls, using the same paint, then you have to figure the ceiling area in your computations. If you're going to use different paint on the ceiling, keep this measurement separate. In either case, you need to know the area of the ceiling.

To find the ceiling area, multiply the length of the room by the width. Notice that *the ceiling area is the same as the floor area!*

Assuming the ceiling is to be a different color, divide the ceiling area by the spread rate for the ceiling paint to find the number of gallons of paint needed for the ceiling.

EXAMPLE: After some discussion, we decided to paint the bedroom walls a light coffee color and the ceiling bone white. The room measures 13 feet 3 inches by 16 feet 7 inches. The height of the ceiling is 8 feet. The coffee-colored wall paint has a spread rate of 450 square feet and the bone paint will cover 500 square feet per gallon. Now let's figure out how much wall and ceiling paint we need.

(1) Round up the room measurements to 14 feet by 17 feet.

(2) Then compute the areas of two adjacent walls by multiplying the length by the height of the room:

14 ft. × 8 ft. = 112 sq. ft.
17 ft. × 8 ft. = 136 sq. ft.
　　　Total = 248 sq. ft.

Since the other two walls have the same area, 248 square feet:

Total area of the walls = 2 × 248 sq. ft. = 496 sq. ft.

(3) Next, calculate the number of gallons of paint needed by dividing the total area of the walls by the spread rate of the paint.

✓✓*HOT TIP*   If a spread rate range is given, use the lower end in this calculation to be on the safe side.

Gallon needed = 496 ÷ 450 = 1.1

This comes to one gallon and one-tenth of a gallon, but it would be best to buy one gallon and an additional *quart* of paint. If the quart is not a custom mix, you can probably return it if you don't open it.

✓✓*HOT TIP*   Save one of the small walls to paint last. That way, when you get to this wall, if it looks like there's enough paint in the gallon can to complete it, go ahead and finish the job. If, however, it looks like you might run out of paint, use the extra quart instead of finishing the gallon can. Paint color varies slightly from can to can. You won't notice the difference on adjacent walls, but will see the color variation if you start a new can in the middle of a wall.

Finally, determine the area of the ceiling.

14 ft. × 17 ft. = 238 sq. ft.

And divide by the spread rate of the ceiling paint (500 square feet per gallon):

238 ÷ 500 = .48

This is slightly less than one-half gallon, so two quarts will do. You can avoid color variation by mixing the two quarts together before starting on the ceiling.

✓*TIP*   Transfer left-over paint into small jars. Try the kind used for canning because they have air-tight seals. Be sure to label the jar with the brand of paint, name/number of the color, base (oil, latex), and which room was painted that color. Using small glass jars for leftover paint takes up less room than old paint cans. They are also more tidy to store and easier to use for touch-ups.

Section 2: How to Make a Round Tablecloth

To make a round tablecloth, you first need to make a square whose side is equal to the diameter of the required circle. Then you draw the circle and cut it out, hem it and, *violá,* you have a round tablecloth.

Let's go through these steps a little more slowly and in greater detail, taking into account some of the geometric concepts involved in circles and squares as well as the sewing suggestions that will make this a truly professional job.

What we have in mind is a floor-length table covering that can be used as a cloth but that is more likely to be used decoratively, perhaps with another cloth draped over it. The following pictures illustrate (A) the table we want to cover, (B) the table covered by the cloth, (C) an "x-ray" view of the covered table, and (D) the table covering spread out flat. As you can see in the last drawing, the cloth is a perfect circle.

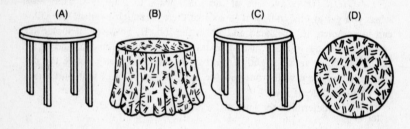

Note that the same principles work for any type of round table top—of any height, with any kind of base or size top.

The first step involves making a square of fabric whose side is the diameter of the circle you need. What exactly is the diameter, and how do you find the diameter of the table covering you want to make?

Here's how

FINDING THE DIAMETER

The diameter of the floor-length round cloth you are going to make will be equal to 2 times the height of the table *plus* the diameter of the round table top *plus* another 2 inches (for the hem).

Step 1 Measure the height of the table from the floor to the top of the table (most dining-type tables are about 29 inches tall) and multiply by 2:

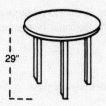

29"

2×29 in. $= 58$ in.

Step 2 Find the diameter of the table top. Drawings A, B, and C are aerial views of the top of a round table. (The dot, "O", represents the center of the circle.)

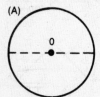

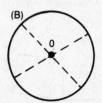

The diameter is the distance of the imaginary line that goes from one point of the circle to another through the center point. It is the length of the longest line you can draw from point to point.

The broken line in Drawing A is the diameter and, as an example, in the table we are covering, it is 36 inches across. Drawing B shows that no matter which way you measure the diameter of this same circle, it is always the same length.

✓*TIP*   Since your real-life table doesn't have the exact center marked for you, it might help in finding the diameter to remember it is the *longest* distance from one point on a circle to another. In Drawing C you can visually compare the length of the diameter with the length of any other line that cuts across the circle but doesn't go through the center. Take a ruler and actually make the measurements so that you don't have to rely on visual data alone.

So, *to find the diameter of the table top, take your tape measure and find the longest distance across the top.* (If the table top is too big for your arm span, fasten the tape measure at one edge of the table and swing the tape along the top until you find the longest length.)

In our example, the diameter measured 36 inches.

Step 3   The diameter of the floor-length table covering is equal to 2 times the height plus the diameter of the top plus 2 inches.

58 in. + 36 in. + 2 in. = 96 in.

Now we have to make a 96-inch square piece of fabric.
Here's how

MAKING A SQUARE

A square is a rectangle with all four sides equal in length and, like all rectangles, it has four right-angle corners. Large and small squares are illustrated below:

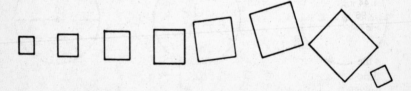

A 96-inch square is a square whose sides measure 96 inches. It is a large square and looks like this:

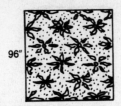

96″

The problem is that, while squares are very simple things, a 96-inch square piece of fabric is a little difficult to make in real life because fabric doesn't usually come in 96-inch widths. If it did, all you would have to do is to measure out a 96-inch length and you'd have the size square you need. Similarly, if fabric came in widths wider than 96 inches, all you would have to do to get a square this size would be to trim the width down to 96 inches, then measure out the required length.

Fabric usually comes in 36-inch, 48-inch, 52-inch, and 60-inch widths. Let's assume that the pattern we want for the table covering comes in a 52-inch width.

To make a 96-inch square from a 52-inch wide fabric, you'll have to add 44 inches to the width:

$$96 - 52 = 44$$

So your square would look like this:

96″

52″   44″

The 44 inch piece of fabric, which is 96″ long, has to be trimmed down from a piece 52″ × 96″. Note that you have to buy two runs of 96″ lengths, or $2 \times 96″ = 192$ inches of fabric (which is 16 feet or 5 yards and 1 foot of material.)

In the illustration above, the dotted line represents where the two pieces of fabric could be joined together. (If your fabric comes in a 48-inch width, you couldn't make a 96-inch square from two runs because you have to allow about one-fourth to one-half inch on each run for sewing the pieces together.)

✓✓*HOT TIP*   If you sew the pieces together like this—

your finished tablecloth will have a seam running across the top. To get the seams on the sides near the bottom where they will be less noticeable, divide the 44-inch piece in half *lengthwise* to get two pieces, each of which measures 22 inches by 96 inches. Then sew one of these pieces to each side of the 52-inch by 96-inch piece, like this:

This will guarantee that no seams will fall on the flat surface of the table top.

Now from the 96-inch square we are going to cut out a circle that has a 96-inch diameter.

Here's how

CUTTING THE CIRCLE

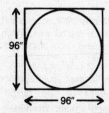

The picture above illustrates what we want to accomplish. The circle we want will touch the square on all four sides. Can you see how this circle has a 96-inch diameter?

In school, to draw a circle you used a compass. The pointy leg left a teeny hole that marked the center of the circle you were drawing, and the distance to the pencil point was called the *radius* of the circle.

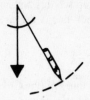

If you recall, the *radius of a circle is equal to one-half the diameter* (or, the diameter was 2 times the radius). So, we need a compass that can draw a circle with a 48-inch radius (96 in. ÷ 2). That's a mighty big compass: 4 feet. It's certainly not an instrument we'd find lying around the house so . . . Let's make a compass.

✓*TIP*   To make a compass, we will need a pin, a long length of string (more than 48 inches), and a piece of chalk or charcoal. Some tape would also be helpful.

(1) Tie the string securely around the chalk and tape it down so it won't slip. Measure out 48 inches of string starting from the knot and mark the length on the string with a pen. Use a pin to mark this spot that will become the ''pointy'' leg of the compass.

(2) Now find the center of the square, which will also be the center of the circle, either by locating the point of intersection of the diagonals (drawing A), or by folding the square in quarters (Drawing B). Mark the center point with the chalk.

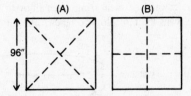

(3) Now spread the cloth out flat on a surface such as the floor and pin the string at the place you marked (48 inches from the end knotted around the chalk) at the exact center of the square. Holding down the center with one hand, carefully extend the string and start drawing the outer edge of the circle. Remember, the string is equal to the radius of the circle, which in our example is 48 inches.

✓✓*HOT TIP*   Try not to either stretch the string or let it sag. Hold it taut, but not tight. Also, it helps if the fabric can be held firmly in place. You might want to borrow another set of hands for this part of the operation.

(4) Once you have outlined the circle, all that remains is cutting it out and hemming it. Use a sharp scissors and cut along the chalk line you have made. Incidentally, the chalk line is the *circumference* of the circle, which is:

$$C = \text{Circumference} = 2\pi R$$

That is, circumference is 2 times pi (the Greek letter) times the radius. Since π equals approximately 3.14 and, in our example, R = 48 inches:

$$
\begin{aligned}
C &= 2 \times 3.14 \times 48 \text{ in.} \\
&= 301.44 \text{ in.} \\
&= 25.12 \text{ ft. or } 8.37 \text{ yds.}
\end{aligned}
$$

It's important to be able to calculate the circumference of the circle if you want to buy trim to edge the cloth with. In our case, you would buy 9 yards of trim. (Trim is sewn on after the hem is made.)

The last step involves sewing the hem.

✓*TIP*   We have left 2 inches, an inch all around, for the hem. Since a round tablecloth is always cut on the bias (that is, on an angle to the weave of the fabric), a "rolled" hem works best. A rolled hem is the type you will find on most handkerchiefs and requires a "hemming" stitch. If you are going to roll the hem, you will have to trim off a little fabric all around first.

You now have a round tablecloth that should cover your table and just reach the floor on all sides. If you want a round tablecloth that doesn't come to the floor, go back to the first step in this section, where we measured the height of the table. Instead of measuring from floor to table top, measure from the point you want the tablecloth to reach, like this:

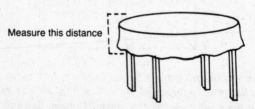

Measure this distance

Double it, add the diameter of the table top plus a couple of inches for the hem, and proceed as we've outlined.

While we were writing this section, we described it to a friend who asked, "Why on earth would you want to make a round tablecloth?" I guess we forgot to start by telling her that we had a round table.

Section 3: . . . With the Fringe on Top: Computing Perimeter

Molding is generally put along the base of walls or along the top edge where the wall and ceiling meet to give a room a more finished look. Edging is usually sewn around the border of area rugs. Round tablecloths, like the one we made in Section 2, often have trim along the bottom. And many people enclose their property by placing a fence around it.

These constructions all involve the concept of *perimeter*.

The perimeter of a flat area is the distance around it. So to calculate, say, how much edging or fencing you need to buy for your home improvement project, you need to know how to compute the perimeter of variously shaped area.

Here's how

FINDING PERIMETERS OF RECTANGLES

To find the perimeter of a rectangle, add the length and width and multiply by two.

This can be summarized by the formula:

$$\text{Perimeter} = 2 \times (L + W)$$

EXAMPLE: Find the perimeter of a rectangle like the one pictured below that is 6 feet wide and 8 feet long.

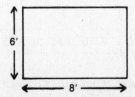

SOLUTION: Perimeter $= 2 \times (L + W)$
$$= 2 \times (8 \text{ ft.} + 6 \text{ ft.})$$
$$= 2 \times 14 \text{ ft.}$$
$$= 28 \text{ ft.}$$

This amount, 28 feet, is the length of molding needed to go completely around the baseboard of a room that is 6 feet wide and 8 feet long. Actually, the required amount of molding would be a bit less, depending on how many feet are taken up by doors and closets. But, when buying molding, edging, fringe, border, or even fencing, always get a bit extra—just in case!

Calculating the perimeter of a *square* is done similarly.
Here's how

FINDING PERIMETER OF A SQUARE

To find the perimeter of a square, multiply the length of a side by 4.

Note that computing the perimeter of a square is really the same as finding the perimeter of a rectangle, because a square is just a rectangle with all sides equal.

EXAMPLE: Find the perimeter of a 7-foot square
SOLUTION: $4 \times 7 \text{ ft.} = 28 \text{ ft.}$

Perimeters of other-shaped flat figures, even odd-shaped ones like the one pictured below, are found in much the same way.
Here's how

FINDING PERIMETERS OF ANY REGION

To find the perimeter of any region or area bounded by straight line segments, add up the lengths of the bounding segments.

EXAMPLE: What is the perimeter of the figure illustrated below?

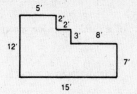

SOLUTION: In this figure we add up the lengths of the edges as we move around the figure, counterclockwise, starting with the left side.
Perimeter = 12 + 15 + 7 + 8 + 3 + 2 + 2 + 5 ft.
= 54 ft.

The distance around a circle is called the *circumference* rather than the perimeter. There is a special formula for finding the circumference (perimeter of a circle).
Here's how

FINDING CIRCUMFERENCE

Use this formula:

$$\text{Circumference} = 2 \times \pi \times R$$

In this formula, R is the *radius* of the circle and the Greek letter π (pi) stands for the number 3.14 (approximately).

EXAMPLE: If a circle has a radius of 7 inches, what is the circumference?
SOLUTION: Circumference = $2 \times 3.14 \times 7$ in.
= 43.96 in. or 44 in. (Rounded off that's about 3 feet and 8 inches.)
EXAMPLE: Find the circumference of a circle that has a radius of 5 inches.

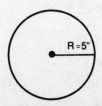

SOLUTION: Circumference $= 2 \times \pi \times R$
$\qquad\qquad\qquad = 2 \times 3.14 \times 5$ in.
$\qquad\qquad\qquad = 31.4$ in.

Since the *diameter* of a circle is twice the radius, another formula for the circumference is:

$$\text{Circumference} = \pi \times D$$

In this formula, D is the diameter of the circle.

EXAMPLE: Find the circumference of a circular running track that has a diameter of 140 yards.

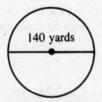

140 yards

SOLUTION: Circumference $= \pi \times D$
$\qquad\qquad\qquad = 3.14 \times 140$ yds.
$\qquad\qquad\qquad = 439.6$ yds.

That's about ¼ mile since one mile $= 1,760$ yards and ¼ mile $= ¼ \times 1,760$ yards $= 440$ yards.

EXAMPLE: Find the amount of fringe needed to finish a round rug that has an 8-foot diameter.
SOLUTION: The amount of fringe needed is equal to the circumference of the rug.
Circumference $= \pi \times D$
$\qquad\qquad\qquad = 3.14 \times 8$ ft.
$\qquad\qquad\qquad = 25.12$ ft.

So, you would need to buy 26 feet of fringe.

12

Figuring
Your Bill

Section 1: It's a Gas! Gas Meters and Gas Rates

When you use gas for cooking (assuming, of course, that you have a gas stove) or for heating, the flow of gas through the gas lines is measured by a meter. Gas meters are usually installed inside your home or apartment, in the basement. Once a month, the gas company sends someone to read your meter in order to determine how much gas you used. If you are not at home and they are unable to read the meter, the utility will estimate your monthly usage based on your past usage patterns. When they next read your meter, they will make corrections in your bill, up or down, if the estimate did not agree with the actual usage.

Gas consumption has traditionally been measured in *cubic feet* and you are billed for the number of *hundreds of cubic feet* you use. Starting in June 1983, Consolidated Edison in New York City changed from a base of one-hundred cubic feet to "therms."

A therm is a measure of the heat content of the gas supplied to you. It is generally a little more than one-hundred cubic feet (100.3 cubic feet) but, for all practical purposes, one therm and one Ccf (as one-hundred cubic feet is abbreviated) can be considered equivalent. Actually, the *heat content* of the gas you get varies slightly from month to month.

Low gas usage would be about 200 cubic feet (2 therms) per month. That's the amount you would use for running the pilot light on your stove and for light cooking. Obviously, gas usage would be far higher if you did a lot of cooking and baking and if you also used gas for heating your house or for making hot water.

You can, if you wish, read the gas meter yourself—we will show you how. But the gas company will not allow you to call in a reading. They always want to read it themselves.

Gas meters are of two types: the older meters are made up of dials; the newer ones are digital and extremely easy to read.

First, we'll discuss how digital gas meters work.

Here's how

DIGITAL GAS METERS

A digital meter looks something like this:

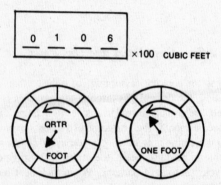

The numbers on the display represent the number of hundreds of cubic feet used since the meter was first installed. The dials let you see how fast you are consuming gas. The small dial on the left makes one complete revolution for every ¼ cubic foot of gas used. The small dial on the right makes one revolution per one cubic foot of gas. Thus, 4 revolutions of the dial on the left correspond to one complete revolution of the dial on the right.

To see how it works, put up a kettle of water to boil and then go and look at the gas meter. You will see the hand on the dial on the left moving rather quickly, while the hand on the dial on the right moves more slowly: 4 revolutions to one. (The digital display dials will probably not move because the smallest unit of measurement there is 100 cubic feet.) If you were to do the same experiment when roasting a turkey in the oven and cooking a pot of soup on the stove, you would see the dials moving more rapidly because of the increased rate of gas consumption.

In the illustration above, the meter reads 106 hundreds of cubic feet. That's actually 10600 or 10,600 cubic feet. One month later the meter looks like:

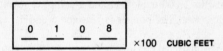

It now reads 108 hundreds of cubic feet. The difference in the digital readings from one month to the next represents the gas usage for that month. The gas usage in our example is: $108 - 106 = 2$ hundreds of cubic feet. Your gas bill is computed by multiplying the monthly usage by the gas rate—plus minimum charges, taxes, and any gas adjustment factor.

Now, let's examine dial gas meters.

Here's how

DIAL GAS METERS

A dial gas meter looks something like this:

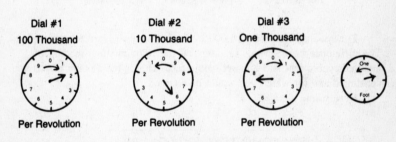

To read these meters, begin with Dial #3. The "one thousand" written above the dial means that one complete revolution of the hand represents 1,000 cubic feet of gas consumption. (Note that the arrow ⌒ on the face of this dial indicates that the hand moves *clockwise*.) Since 10 hundreds makes one thousand, the numbers 1–10 represent 100, 200, 300, etc. cubic feet of gas consumption. The hand is pointing between 7 and 8, so this dial indicates 700+ cubic feet.

Now go to Dial #1, the first dial on the left. One revolution of the pointer represents 100,000 cubic feet and each number on the dial stands for

10,000 cubic feet (since $10 \times 10,000 = 100,000$). The arrow ⌒ indicates that the rotation direction is clockwise. Since the hand is between 1 and 2, this dial reads 10,000+ cubic feet.

Dial #2 moves *counterclockwise* as indicated by the arrow ⌒ . Each number represents 1,000 cubic feet and a complete revolution is $10 \times 1,000$ or 10,000 cubic feet. The pointer is between 5 and 6, representing 5,000+ cubic feet.

The three dials taken together read 15,700 cubic feet (10,000 + 5,000 + 700). (The small unnumbered dial on the right is not part of the gas company's meter reading; it helps you see how quickly you are consuming gas, with one revolution representing one cubic foot of gas consumption.) Notice that although the three dials read 15,700 cubic feet, the reading would be recorded as 157, meaning 157 hundreds cubic feet.

Suppose that at the end of one month the meter looks like this:

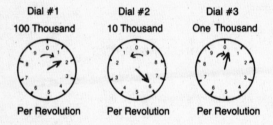

The reading would be 16,000 cubic feet or 160 hundreds cubic feet. The difference between this reading and the prior month's reading of 157 hundreds cubic feet is 3 hundred cubic feet. This is the amount of gas you consumed and for which you would be charged.

Here's how

COMPUTING YOUR GAS BILL

A local utility charges its residential customers a minimum of $7.13 a month, which includes charges for billing, meter reading, and other customer services, and also covers the first 3 therms of gas consumption. If you don't use gas for heating, you are then charged 68.5¢ per therm for all therms over 3. If, however, you do have gas heat, your additional therms are less costly (since the utility figures that you'll be using so many more of them): 66.81¢ per therm for from 3 to 3,000 therms and 60¢ per therm for therms over 3,000 per month.

If, as in our example, you used only 3 therms (or 3 hundred cubic feet), you would pay only the minimum usage charge (plus the other charges described below). If you used 5 therms you would pay:

$7.13 (which includes the cost of 3 therms)
+ $1.37 (2 × 68.5¢)
    $8.50

In addition to the minimum charge and excess usage charges, this power company also has a 5¢ per therm adjustment factor. Some gas companies now use adjustment factors to compensate for changes in the price they pay for gas. The adjustment factor is the way they pass along increases or decreases in price to the consumer.

If we used 5 therms at a total of $8.50, and if there had been a pass-along adjustment reflecting increased gas costs, the charge for your gas consumption that month would be:

$$\$8.50 + (3 \times 5¢) = \$8.65$$

That's not all. Calculated into your bill, but not shown, is a 6.86% gross receipts tax. There's also a 4% sales tax if you are a residential customer. That's an additional 94¢ (10.86% × $8.65):

$$8.65 + .94 = \$9.59$$

So, using only a minimal amount of gas, your bill quickly adds up to almost $10. It's a gas, all right!

Section 2: Electrical Charges or How Your Electrical Bill Is Computed

When you turn on a light switch, electrons immediately begin to move through the wire, generating electricity. The *rate of flow* of electrons through a wire is measured in units called *amps*, which is short for amperes, after the French physicist, André Marie Ampère (1775–1836), a principal architect of electromagnetic theory. The symbol i is generally used for amps or *current*.

All electrical devices exhibit *resistance* to the flow of current. This resistance is similar to the friction that is present in mechanical devices. The energy that is dissipated in overcoming resistance (friction) appears as heat. Electrical resistance is measured in units called *ohms*, named after the German physicist George S. Ohm (1787–1854). The usual symbol for resistance or ohms is r.

It takes work to overcome resistance, and *the volt* (named for the Italian physicist Alessandro Volta (1745–1827), *is the unit of electromotive force required to move a current of one amp through a resistance of one ohm*.

Voltage, current, and resistance are related through Ohm's law:

$$v = ir$$

In this law, i is the current in amps, r is the resistance in ohms, and v is the voltage in volts.

The "electricity" we pay for, however, is electrical *power*.
Here's how

WATTS AND KILOWATTS

Although we tend to think of power in informal terms, it has a precise meaning in physics and to the "power" companies. *Power is defined as the rate at which electrical energy is transferred.* Power is measured in *watts* to acknowledge the Scottish inventor, James Watt (1736–1819).

Watts (power) is the product of voltage and current measured in volts and amps, respectively. That is:

$$p = vi$$

In this formula p stands for power in watts.

We will see in a moment that it doesn't take much to use one watt of electrical power. Since watts are used rather quickly, electrical power is generally quoted in *kilowatts:*

1 kilowatt = 1,000 watts

Actually, we don't pay for electrical power per se, but for how long that power is used. After all, it is clear that we should be charged more for using one kilowatt of power for an hour rather than for a second. So electric rates are based on the cost per kilowatt-hour (kwh). *One kilowatt-hour means that 1 kilowatt (1,000 watts) is transferred continuously for a period of one hour*.

Now let's examine what this means in practical terms.

Consider light bulbs. A typical light bulb might be rated as 100 watts. It takes 100 watts of electrical power to light the bulb! Since 1,000 watts equals 1 kilowatt, this means that it takes 0.1 (¹⁄₁₀) kilowatt of power to light the bulb. If the lamp is on all evening, say from 6:00 PM to 11:00 PM, the light burns for 5 hours. Therefore, you will be charged for 0.1 kilo-watt × 5 hours = 0.5 kilowatt-hours of electricity.

Electricity rates range from about 5 to about 15¢/kwh, depending upon the region of the country in which you live and the amount of power you use. In one locality, for example, the residential rates are 13.281¢/kwh for the first 250 kwh's, then 12.756¢/kwh for kilowatt-hours in excess of 250. These are the winter rates, when you pay less per kilowatt hour for using more electricity. There is an inverted rate for the summer months (from June 1 through September 30): 13.2814¢/kwh for the first 250 kwh's, then 14.256¢/kwh for anything over 250 kwh's. In the summer when more power is in demand (mostly to run air conditioners), you pay a penalty of a higher per kilowatt-hour rate for using more power.

Let's suppose you are paying a high rate of 15¢/kwh. Keeping the 100-watt bulb burning for 5 hours will cost you:

.5 kwh × 15¢ = 7.5¢

That doesn't sound like much for having the lamp on all evening. But wait!

EXAMPLE: During a typical evening, you may have five lights on, two of which are rated 60 W (watts), another two at 100 W, and the final one at 150 W. You also have the television set on and it is rated at 120 W. (The wattage of an electrical device is generally printed on the bottom or at the back near the power cord.) Suppose all these devices are on for 5 hours. Let's figure what the cost will be at the above rate of 13.281¢/kwh.

SOLUTION:

Step 1. First, you compute the total number of watts that are used by adding together the wattage of all the equipment:

Total watts = (2 × 60) + (2 × 100) + 150 + 120 = 590

Step 2. Now compute the total number of kilowatts by dividing by 1,000:

Total kilowatts = 590 ÷ 1,000
= 0.59

Step 3. Then compute the number of kilowatt-hours by multiplying the number of kilowatts by 5 (hours):

Total kilowatt-hours = 5 × 0.59
= 2.95

Step 4. Finally, compute the cost by multiplying the number of kilowatt hours by 13.281¢:

13.281¢/kwh × 2.95kwh = 39.18¢

That still doesn't sound too bad. But suppose you do the same thing every day, 30 days a month. Now the cost for the five lamps and the TV set for 5 hours per night is 30 × 39.18¢ or $11.75—and this doesn't even begin to account for all the other electrical equipment you probably use.

Electrical appliances that are *very* heavy consumers of electrical power are toaster ovens, hair dryers, air conditioners, electrical stoves, irons, and refrigerators. For example, a toaster oven is rated at 1350 W and an iron at 1100 W. That's 2 to 3 times more wattage than those five light bulbs and TV set combined!

How does the power company keep track of the number of kilowatt-hours of electrical power you use?

Here's how

ELECTRIC METERS

The electric meter, installed by the power company and usually mounted outside the house or in the basement, constantly records your use of kilowatt-hours.

The newer electric meters are digital, in which case there are no complexities in reading the display: what you see is what you get. The older meters are similar to the gas meters discussed in the previous section. If you have one of the older-type meter boxes, it will probably have a four-dial display like this:

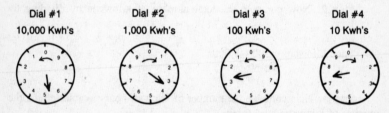

One complete revolution of the hand in Dial #1 represents 10,000 kwh's. Since the hand in the example is between 5 and 6, it represents between 5,000 and 6,000 kwh's. We read this as 5,000.

One complete revolution of the hand in the second dial (#2) represents 1,000 kwh's, so each digit represents hundreds of kwh's. In our example, the hand is between 3 and 4—between 300 and 400 kwh's. (If the hand had gone full circle, it would be 1,000 kwh's). We read the second dial as 300. Thus far we have used 5,300 kwh's.

One complete revolution of Dial #3 represents 100, so each digit represents 10. Since the hand is between 2 and 3, this means between 20 and 30. We read it as 20, and we now have 5,320 kwh's.

The last dial (#4) represents ones, since one complete revolution is 10 kwh's. Since the hand is between 7 and 8, we read it as 7 kwh's. Taken together, the 4 dials read 5,327 kwh's. (Notice that in reading the meter we always read the *lower* of the numbers between which the hand falls.)

The number you read on the meter is not the number of kilowatt-hours you used this month, fortunately, but rather, the total number of kilowatt-hours that have been used since the meter was first installed. The meter is read each month, and the difference in readings from one month to the next is the amount of electricity you have used.

Your total bill for the month includes:

1. A monthly customer charge that covers the utility's cost of reading your meter, billing you, and other customer services.
2. Taxes, including local sales tax.
3. The amount of electricity you have used multiplied by the rate that gives you—pun intended—your monthly charge.

Section 3: Buddy, Can You Spare a Nickel? Dime? Quarter?: Figuring Your Phone Costs

Once upon a time, before deregulation of the telephone industry, it was all comparatively simple. A telephone call cost a dime, double the price of only a short while before that. For a set fee you had a wired phone and were permitted a certain number of local calls. Additional charges got you additional local and long-distance calls. And, while you never quite understood "message units," you got *one* bill each month, from *one* telephone company, most typically a subsidiary of the Bell system. In those days, repairs were generally free or easy to come by and, if you wanted a new instrument, your choice was limited to about five basic models in a narrow range of colors.

Today it's all different. A federal court decision broke the AT&T monopoly, encouraging competition. One early consequence of "deregulation," which went into effect in 1984, has been an increase in the cost of local calls, the result of an attempt to bring the price of basic telephone service closer to its actual cost. Now there's also a late payment charge on unpaid balances. Long-distance rates, however, have tended to stay the same or, in some cases, to decrease. We now have a choice of long-distance carriers (and must learn to be better informed consumers) and a staggering array of new phone instruments to choose from.

Let's start by looking at the price of a local call made from your home telephone.

Here's how

TYPES OF HOME SERVICE

The cost of a local telephone call depends on a number of factors— whether you're using a pay phone, a home phone or an office phone, whether you've requested operator assistance, the type of service you have purchased, and so on.

There is a basic monthly service fee for every phone that consists of: (a) the cost of an exchange access line from the phone company's central office to your home and (b) a line charge ordered by the federal government (Federal Communications Commission).

The *exchange access line* cost depends on the type of service you have. Our company offers its customers two choices:

1. *Life Line Service,* which is the lowest-priced individual phone line service. This service is designed for eligible low-income customers. The cost is $1.04 per month for the exchange access line (all prices are quoted without taxes and other local surcharges). Users of this service get no monthly allowance for outgoing calls—it's designed principally for people who want to receive calls but who make very few outgoing ones.
2. *Residence Message Rate Service,* for which the basic monthly exchange access line charge is $6.60. With this service, calls within "Home Region" are untimed and other "local" calls bear a per minute charge.

There is an additional $3.50 monthly subscriber line charge for every line going into the 90 million households that have phones.

These are basic charges, exclusive of the lease or purchase cost of equipment. The cost of a phone call is based on distance, when (the time of day and day of week) you make the call, and the length of your conversation.

Here's how

LOCAL CALLS

Each telephone company has a defined "regional calling area." A call to an exchange within this area is handled by the local phone company. Our telephone company defines the regional calling area as New York City's 5 boroughs and all or parts of several surrounding counties. In the front of your telephone *Directory*'s "white pages," you'll find the definition of your *regional calling area*.

There are two types of regional calls: *local area* calls to your own or to nearby exchanges, and *extended area* calls, which include exchanges beyond the primary area but within your regional calling area. Local area calls may be *timed* or *untimed*, depending on your type of service, but all extended area calls are *timed*, meaning you get billed for an initial period (of one minute) and for each additional minute that you're on the line. (Calls beyond your regional calling area are considered to be long-distance calls. We'll talk more about these regional areas later.)

Your telephone *Directory* has tables that help you determine your local and extended area exchanges. Another table in the *Directory* lists the charges for each primary and extended area call. Extended area calls are all timed, and all are at higher rates. Our telephone company's *day period rates* for local area calls are 10.6¢ for untimed and basic budget service calls.

These charges can mount up quickly. One way to keep costs down is to try to take advantage of discounted rates.

Here's how

DISCOUNTED RATES

Telephone companies have different rates depending upon whether you make your calls during peak business periods (expensive) or during slack, off-hours (cheaper). The discounted rates are also shown clearly by your telephone company.

Our phone company has *day period rates* that apply weekdays from 8:00 AM to 9:00 PM, except for five holidays (Christmas, New Year's, Thanksgiving, Independence Day, and Labor Day) when evening rates apply from 8:00 AM to 11 PM and night rates apply from 11:00 PM to 8:00 AM. Day period rates are most costly.

Evening period rates, which are in effect from 9:00 PM to 11:00 PM, Mondays–Fridays, are discounted by 40 percent. (See Chapter 5, Section 2 for a complete discussion of discounts and an explanation of how to compute them.)

✓*TIP*   The least expensive time to call, if you can, is when *night period rates* are in effect; there's a 65 percent discount on the cost of calls made Monday to Friday from 11:00 PM to 8:00 AM, all day and night Saturday, and Sunday from 8:00 AM to 5:00 PM and from 11:00 PM to 8:00 AM.

The phone company suggests that the "most accurate way to figure the cost of an evening or night call is to use the full week day rate for the total time, then subtract the 40 percent or the 65 percent discount and round to the lower penny."

Let's compare the cost of the same call made when the three different rates are in effect.

Suppose we had timed service and made a local area call for which we are charged the rate of 14.6¢ for the first minute and 4.7¢ for each additional minute. These are day period rates. If we talked for 17 minutes, the charge for this call (exclusive of taxes and local surcharges) would be:

$$14.6¢ + (17 \times 4.7¢) = 94.5¢$$

If, however, we made this call between 9:00 PM and 11:00 PM when evening rates were in effect, we would get a 40 percent discount. The best way to compute the discount is to: (a) compute the total day period cost of the call, (b) calculate 40% and (c) subtract the 40% from the total day cost, rounding to the lower penny:

(a) 94.5¢
(b) $0.40 \times 94.5¢ = 37.8¢$
(c) $94.5¢ - 37.8¢ = 56.7¢$ or 56¢ rounded to the lower penny

If we made the same call at a 65 percent discount during the night hours, it would cost:

(b) $0.65 \times 94.5¢ = 61.4¢$
(c) $94.5¢ - 61.4¢ = 33.1¢$ or 33¢ rounded to the lower penny

What happens if you start your call in one rate period and end it in the next one? Charges are billed at the rate that applies for each period. So if you started your call Monday at 10:50 PM and talked for 17 minutes, you'd be billed overtime at the evening period rate for 9 minutes and at the night period rate for the next 8 minutes:

(a) $14.6¢ + (9 \times 4.7¢) = 56.9¢$
(b) $0.40 \times 56.9¢ = 22.8¢$
(c) $56.9¢ - 22.8¢ = 34.1¢$

+

(a) $8 \times 4.7¢ = 37.6¢$
(b) $0.65 \times 37.6¢ = 24.4¢$
(c) $37.6¢ - 24.4¢ = 13.2¢$

$34.1¢ + 13.2¢ = 47.3¢$ or 47¢

In making any kind of call, even a local one, you pay extra for such services as an operator-assisted person-to-person call ($3.49 by our telephone company!), an operator-assisted station-to-station call ($1.58), a calling card call, and for billing the call to a third number. Discount rates never apply to these extra charges but continue to apply to the timed charges.

In the olden days, before deregulation, you had local calls and long-distance calls. What happened to long-distance?

Here's how

REGIONAL CALLING AREAS

A *Regional Calling Area* is a geographic area within which the local telephone company provides local and long-distance services, plus access to the telephone network. What this means is that for some defined distance beyond your local calling area, your local telephone company provides you with long-distance service.

For example, the New York Telephone Company has seven Regional Calling Areas. We live in the "New York Metropolitan" one that covers our local calling area plus some other counties as well as parts of other states. According to our *Directory,* a call to part of Connecticut is a long-distance call handled by our local phone company. But a long-distance call to another part of that state would be handled by the company we have selected as our long distance carrier—AT&T, SPRINT, or MCI, to name the three largest competitors. This long-distance call is billed separately by the long-distance company and, in the case of companies other than AT&T, may involve separate dialing instructions.

As we said before, it's least expensive to make a direct dial long-distance call, whether within or outside your Regional Calling Area and irrespective of which long-distance service you have bought. They all charge extra for extra services (like operator assistance), although the basic calling charge reflects distance, the time of day and day of the week you call, and the length of your conversation.

The *Directory* lists some sample rates for long-distance calls within our Regional Calling Area. Shown are full day rates (between 8:00 AM and 5:00

PM), 40 percent evening discounted rates (from 5:00 PM to 11:00 PM) and 65 percent night and weekend discounts from 11:00 PM to 8:00 AM and Saturdays and Sundays. Charges are based on the rates in effect at the place the call originates, and our telephone company has an initial one-minute period with each additional minute charged as overtime for long-distance calls within the Regional Calling Area.

Our local phone company also offers a package that would be very attractive if we were planning to make several long-distance calls per month within our Regional Calling Area—such as to a child attending a state college or to a special friend who has a job upstate. Check with your phone company to see whether they have similar deals.

Long-distance calls to other Regional Calling Areas, even if those are within your local company's area, are not provided by your local company but by long-distance carriers. Rates for these types of long-distance calls are determined by the companies that provide these services. When examining rates, remember that you are comparison shopping; you will want the least expensive service with the fewest "extra" charges, smallest monthly fees, and best package deals to places you call frequently. But you also want reliable service and ease of dialing. Look for a company that has applied for "equal access!"

Until deregulation, all customers using traditional long-distance dialing procedures (a "1," where required, followed by the area code and phone number) had their calls handled by AT&T. Currently, local companies are modernizing their equipment and facilities to allow all long-distance companies that request it equal access to these traditional dialing procedures so their customers will no longer have to use more cumbersome dialing instructions. Shortly, your long distance company will have equal access privileges; it will then probably notify you of any changes in dialing procedures.

How does deregulation affect your bill?

Here's how

PHONE BILLS

In the past, for many residential customers who had chosen AT&T as their long-distance carrier, there was only one bill with the long-distance carrier charges clearly separated. However, now anyone using AT&T, SPRINT, or MCI, for example, receives a monthly bill from them as well as one from the local company.

Our phone bill is several pages long. Page 1 consists of a summary statement that includes the total amount of the last bill and the payments applied to it. It then shows the prior unpaid balance, if any, and the 1.5 percent per month (that's 18% annually) late payment charge.

Also on the summary are AT&T communications' charges (long-distance), and the due date of the bill after which any unpaid balance is subject to the interest penalty. The *total amount due,* incorporating all of the charges mentioned above, is clearly indicated.

What you should remember when examining your bill is that the charges for local service and equipment are *billed one month in advance,* while local usage, long-distance calls, and any installation and/or repair charges are shown after the fact.

Let's turn to the back-up pages.

If you subscribe to AT&T for long-distance, the long-distance calls are itemized in the "AT&T Communications" section of the bill. You get a printout of the date and time of each call, the location and number called, an indication of whether day, evening, or night rates applied to that call, the length of the call (in minutes), and the amount of each call. Also shown is the total charge for all itemized calls.

There may also be a page that details the current charges by your local telephone company. First, there's the "monthly service" charge, which includes the monthly fee for the type of service you own (including the access line charge). Also included by our telephone company is a $3.50 subscriber line charge. Your local phone bill may contain this or a variety of other charges and/or credits.

Local usage totals appear next and should be fairly easy to interpret. Our bill shows the details of local usage for the local calling areas, together with the rates. Last month, for example, we made 63 calls to our home area. The number of calls we made to each of the other parts of our regional area are also shown, as is the number of calls made during each time interval when different discounted rates were in effect. Total amounts are given for all the calls to each calling area and, these totals together with any local surcharges, amount to your "total local usage charge."

Knowing what you know now, if you take the time to examine your bill, you'll find it's pretty straightforward but, once upon a time, long ago. . . .

Section 4: Keep Your Cool: Btu's and Air Conditioners

Buying an air conditioner, like buying any other major appliance, requires careful consideration. Not only is there style and price to worry about, but there's also Btu's and EER's.

You don't have to understand what a Btu is to buy an air conditioner or to use one, and understanding them won't keep you any cooler in the summer. But we think the concept behind Btu's is interesting and will give you a better feel for what air conditioners do.

Here's how

BTU's

Although on an intuitive level we all know what heat is, it was only about 150 years ago that scientists came to recognize that heat is a form of energy. One Btu *(British thermal unit)* is the quantity of heat (energy) that enters (or leaves) one pound of water when the temperature of the water is raised from 63°F to 64°F (or lowered from 64°F to 63°F).

In the metric system, the calorie is a measure of heat. One calorie is the quantity of heat (energy) that enters (or leaves) one gram of water when the temperature of this water is raised from 14.5°C to 15.5°C.*

A little computation would show that:

1 Btu = 252 calories

The output of air conditioners is rated in Btu's per hour because they must cool—lower the temperature—of your room(s). But how do you know how many Btu's are necessary for the size room(s) you wish to cool?

Here's how

THE PROPER AIR CONDITIONER SIZE

Actually, in this book we are not able to provide all the information needed to determine the right size air conditioner for you. But we will tell you generally what you need to know and where to find additional information. We'll also give you a rule of thumb used by our local appliance dealer to get an approximation of your Btu needs.

Determining the proper air conditioner capacity warrants study. The size you need, measured in Btu's, is based on the size of your room, the height of the ceiling, the number of doorways and arches, the number of

---

*Reference temperatures such as 63°F or 14.5°C are used in definitions because the amount of heat needed for a one-degree rise in temperature varies a bit with temperature.

windows and the directions in which they face, the type of construction of your walls, whether you're cooling the top floor of the building or a lower floor, the electrical equipment you use, and other minor factors.

✓✓*HOT TIP* You can come up with a rough estimate by *multiplying the square footage (area) by 30 for rooms on lower floors* and *multiplying the square footage (area) by 35 for rooms on the top floor* of the building. Thus, if your first-floor room measures 12 feet × 16 feet:

(1) Compute the area (see Chapter 11, Section 1):
    12 ft. × 16 ft. = 192 sq. ft.
(2) Multiply the area by 30: 192 × 30 = 5,760

According to this guide, your air conditioner should be about 5,800–6,000 Btu's.

But this is only an approximation. It's worth the trouble to more precisely determine the right size. The Association of Home Appliance Manufacturers (AHAM) has developed a form that helps you to compute the Btu-per-hour capacity needed to properly cool your room. Ask your local appliance dealer for this form or to help you obtain a copy. Or write directly to the AHAM. Here's the address and phone number:

Association of Home Appliance Manufacturers
20 N. Wacker Drive
Chicago, IL 60606
(312) 984-5800

We used the form and found it to be easy to follow and remarkably helpful. It saved us considerable money because it turned out we didn't need as large an air conditioning unit as we thought we did. An air conditioner that is too large serves no purpose and will cost you more money to purchase and to run. On the other hand, an air conditioner that is too small will never cool your room adequately—except maybe in the late fall or winter when it's already cool enough.

Another factor to consider in buying an air conditioner is its Energy Efficiency Ratio (EER).

Here's how

EER

The Energy Efficiency Ratio (EER) is *the ratio of net cooling capacity, measured in Btu's per hour, to the total rate of electrical input, measured in watts, under designated operating conditions.* This ratio is indicative of how efficiently the air conditioner will use electricity. The higher the EER, the more efficient the air conditioner and the less it costs to operate. EER's range from 4.0 to 12.0. Low efficiency units are rated 6.5 or less; the efficient ones have an EER of 8.5 or more. A unit with an EER of 4.0, for example, will cost about 3 times as much to operate as a model with an EER rating of 12.

In our local appliance store, the tag on one typical air conditioner said that it costs 13.5¢/hour in electricity to operate a 6,000 Btu air conditioner unit with an EER rating of 7.6. In contrast, it costs 11¢/hour to operate a more efficient (EER = 9.5) model of the same size by the same manufacturer. If you run your air conditioner for about 700 hours per year, the average number of operating hours for our area of the country, you'll save $17.50 a year by using the more efficient unit. There will be bigger savings if you have a larger-size air conditioner.

Generally, high efficiency air conditioners cost more than lower efficiency units of the same size. We priced two models recently, each of 6,000 Btu's. The one with the EER rating of 9.5 had a list price of $424, while the model with the lower EER rating (7.6) listed for $335. Based on purchase price alone, the higher cost of the energy-efficient air conditioner would take five years of average use to make up. If you expect your air conditioner to last longer, you will save money in the long run, and you'll be making a contribution to the conservation of our energy resources.

Whatever your choice, happy shopping and have a cool and efficient summer!

13

Room to Grow: Personal Computers

In this "computer age," it's hard to resist the push and pull of the deluge of advertisements and articles about personal computers. There are dozens of magazines devoted exclusively to them; business magazines are full of ads for computers and software programs; even your daily newspaper may have a regular computer column in addition to scores of articles about computers. They're constantly mentioned on television and radio and in nontechnical publications of all kinds. Schools clamor for more computers, and computer literacy is the talk of the day.

Essentially, personal computers (PCs) or home computers process sequences of coded instructions extremely rapidly. Their power and versatility stem from this ability to quickly and accurately follow commands. These commands are generally invisible to most computer users who simply buy ready-to-use programs for particular applications.

Rapid technological advancements continue to result in ever more dramatic increases in both processing speed and in the amount of information (memory) computers can retain. This extraordinary increase in computing power is good news for consumers. Programs have become more sophisticated in terms of their capabilities and, at the same time, have become easier to use. There is more good news. While quality and power continue to improve, prices of computing equipment (hardware) and application programs (software) continue to decline.

In our opinion, not everyone needs a computer. But we do think that everyone should know a little about what they can do. To that end, this chapter covers some computer basics, to help you assess whether or not you really need a computer and to provide some guidelines on what to look for if and when you decide to buy one.

Here's how

WHAT COMPUTERS DO

The following are but a few of the uses of home computers, although not necessarily in the order of popularity:

- Word processing
- Data manipulation
- Financial recordkeeping/check writing
- Personal appointment organizer
- Spreadsheet analysis
- Income tax preparation
- Video games
- Programming
- Telecommunications

We will discuss these applications one by one.

Word Processing

Word processing is typing—but much, much better. It is an understatement to say that word processors are as much of an improvement over the old manual typewriter as a food processor is over a mortar and pestle.

When we speak of word processing capability, we really mean that a small (about 3½ inches by 3½ inches, measured externally) plastic encased disk (or disks) with tens (or hundreds) of thousands of instructions, collectively called a *word processing program*, must be inserted into a slot in the computer (the external disk drive) so that its instructions can be stored on the computer's internal "hard" disk drive. Once this has been done, the disks are no longer needed and the computer can "learn" how to do word processing at any time you like by simply reading the stored instructions.

Doing word processing means that a person types at a computer keyboard and the words appear on a computer screen or *monitor*. The text remains in the computer's internal memory until it is stored in a "file" on the hard disk or on another small disk that is placed in the disk drive. When the person wishes, she can instruct the computer to read the file and transmit the information on it to a *printer*, where the information will be printed automatically.

The advantage to typing on a computer screen is that you can make corrections quickly and easily *before your work is actually typed (printed)!* Have you ever had to retype an entire page because you left out a word or a line or paragraph? With a word processor, that will never happen again. Every imaginable change can be made on the screen. Here are just a few of the things you can do, and it doesn't take terribly long to learn how:

- Type over and correct errors
- Insert or delete words, lines, or paragraphs
- Interchange the order of words, sentences, or paragraphs
- Number pages automatically
- Change from single-space to double-space and vice versa
- Rearrange margins, making them narrower or wider
- Center text automatically
- Create special effects, such as boldface and underline
- Type subscripts and superscripts
- Automatically find and correct spelling errors

Many excellent word processing programs are available for all kinds of personal computers. They are regularly reviewed in computer magazines. Your local computer store can also give you some idea of the options you have.

The more sophisticated word processing programs go far beyond the capabilities listed above. Some programs enable you to control the type and size of font so that page headings can be printed larger than the body of the text, footnotes can be printed smaller, and italics and script can be used wherever you wish. You also may be able to insert diagrams, graphs, and drawings into any part of any document you create. In fact, you can do your own "desktop publishing" and create printed material that looks as if it was done by a professional publisher.

Does it pay to buy a computer *just* to do word processing? If you type 5 to 10 letters a week or an occasional term paper, probably not. But if you do a great deal of writing or typing, it's a remarkably worthwhile investment. It can result in a 25% to 50% increase in speed of production, mainly because of the ease with which errors can be corrected. And, for some reason, it seems to help many people compose: for them, looking at a blank screen is not as intimidating as staring at a blank piece of paper.

Data Manipulation

Data manipulation is *data base management,* which is another way of saying that PC's have the capability to organize all types of data. At home and on the job, people collect information (data) and keep it in files, notebooks, and on index cards. A data base management program consists of thousands of instructions to the computer that enable it to act as filing cabinet, index card file, or telephone directory.

Data entered into the computer and handled by a data base management program is stored, in turn, on a disk (or disks) in a form that makes it very easy to retrieve. But a data base program enables you to do much more than just store and retrieve information. Here are some of the other possibilities:

- Randomly entered information can be quickly sorted alphabetically.
- In just a few moments, a computer file of addresses and telephone numbers can be sorted by city, state, or zip code.
- A file of employee records can be almost instantly sorted by date of employment or by any other characteristic, if it is one of the items in the employee records.
- A checkbook file can print out and total all checks issued by category, if the category was part of the file.
- An inventory file can be used to print out a report of the total purchases made, by category.
- An index file of authors, books and articles on a variety of topics can be rapidly searched. For example, all works by a single author or all articles on a specific topic could be found and listed almost instantaneously.

Do you need a data base management program? Probably not, unless you work with very large quantities of data like those involved in a small business or a major research project. Otherwise, it's simply not worth the effort of setting up a system that requires that you enter (by hand) every single letter and number into the computer file. If you have a file of, let's say, 100 addresses and phone numbers, you would do just as well keeping them in an address book, or in a card file or rolodex. But if you have 10,000 addresses and phone numbers and do mass mailings, you'd be foolish not to set up a data base system.

There are many excellent ones available either as stand-alone products or as part of integrated packages that include spreadsheet programs (which we discuss next). Read the reviews in computer magazines and consult your computer dealer for advice on which data base management program to buy.

Financial Recordkeeping/Check Writing

In contrast to sophisticated data base management programs, there are scaled down programs that do highly specialized tasks extremely well. Do you still have trouble balancing your checkbook or keeping expense records for income tax purposes? If you already have a computer, you may want to consider a program that helps you to organize all of your financial records.

Here is a partial list of what such programs (that cost less than $50) can do:

- Keep track of deposits, cash withdrawals, and checks
- Keep a running total of expenses by categories such as entertainment, books, and travel
- Simultaneously manage a checking account, savings account, and investment account
- Print customized expense and income reports
- Print customized checks and enter them in a register

Personally, we have found that such a program is extremely useful. It also takes some of the pain out of recordkeeping.

Personal Appointment Organizer

Because we use our PCs every day, we have found it convenient to have our appointments appear on the screen when we turn on the computer in the morning. Our appointment program (which costs less than $100) makes the computer look like a pocket diary. We can look at appointments one day or one week at a time and even have a little bell remind us that we have to attend a meeting.

Here is partial list of some of the other items that are included in our appointment program:

- A To Do List section that can be used to remind us of what needs to be done each day.
- A Note section that allows us to jot down random thoughts as they occur.
- A Telephone Directory that provides a convenient place to record names and addresses, just as we would in an ordinary directory.

Spreadsheet Analysis

A spreadsheet program literally creates a *spreadsheet* on your monitor, so that your computer screen looks like *ledger paper*. Anything you could do in a ledger can now be done in the computer.

One of the more remarkable uses of a spreadsheet program is in making cash flow projections. If you own your own business, for example, you can put all your income and expenditures on the computer spreadsheet. With just a few keystrokes you can factor in a 10% increase in costs (as an illustration) and see the ripple effect on your balance sheet 6 months or one year or more later.

Another use of the spreadsheet program is to assist you in keeping your tax records. The advantage is that all columns and rows can be instantly and accurately tallied and, if you wish, the results passed on to a tax preparation program. Remember, though, that setting up such a spreadsheet takes a lot of work. Every number must be entered from the keyboard. There is no other way to do it.

It's hard to imagine how the computer's spreadsheet capability could be helpful to salaried people who have one major source of income and few allowable deductions. On the other hand, spreadsheet analysis programs can be extremely useful to businesses that do or would like to do cash flow projections or scenario analysis.

Income Tax Preparation

If you like the idea of doing your own taxes, a tax preparation program may be just what you need. These programs are updated annually and take account of current changes in the tax laws. Again, it is probably not economical to buy a computer just to do tax preparation. An accountant would be far less costly.

In case you decide to try to use a tax preparation program (which, like many specialized applications, costs less than $50), we offer a word of advice. The first year that you use the program, be sure to compare your results with the results that you (or your accountant) obtain by manually filling out the forms. If the results agree, you can feel more confident in the program and in your ability to use it. In the second year you use the (updated) program, you are likely to be safe in trusting the program's results.

Video Games

It would not be unusual if your first introduction to a PC was through a video game. Such games are fast, exciting, absorbing, and fun. It's easy to become "hooked." If you have never tried one, visit your local video arcade and try some out. At first, you may feel clumsy as you try to manipulate the figures on the screen with the *joystick* (the handle with the control buttons). It does require some good hand-eye and motor coordination. But after some time, and lots of quarters, you may find yourself eagerly trying to beat your own score.

Of course, not everyone enjoys games, and a game that appeals to one person may not appeal to another. However, if you generally enjoy this sort of thing, we have a feeling you could get the video game bug. There are games written for just about every PC.

Should you spend between $1,000 and $3,000 on a computer *just* to play games? That is probably not necessary because very popular game programs such as Nintendo can plug right into your TV set. But, if you are buying a computer anyway, a game or two just might be a fun addition.

Programming

Traditionally, most people associate programming with computers. But programming is not the primary use of home computers. As we have seen, the most common home computer applications are already *pre-programmed* for you.

Of course, programming a home computer yourself is not only possible, but some people also find it challenging and very interesting. It allows them to "customize" their computer environment in a way that no pre-packaged program can do. And programming yourself has the added benefit of giving you more insight into how the computer actually "thinks."

You communicate with the computer through a programming language, such as BASIC, PASCAL, Fortran, or C. Commands (instructions to the computer) are written in the syntax of the language with which you are working. These commands are then translated into the machine language (which consists of strings of zeros and ones) that the computer understands.

Most home computers understand the language BASIC. If you want to work with another language, it is necessary to buy another program, called a *compiler,* which translates the program in this language into machine language.

Telecommunications

It's now possible for your computer to "talk" to other computers through the telephone lines by means of a device called a *modem.* Signals from your computer are transformed into sounds (beeps) that are then transmitted along phone lines by the modem and picked up by another modem at the host computer. The sounds are then transformed back into computer signals that the other computer understands. Different modems are designed for different computers.

Many banks now have a special service that, for a fee, lets you view your bank records and transfer money via your home computer. There are also services you can buy that allow you to access the latest stock market activity on your computer screen. You can see the latest airline schedule and book flights through your home computer. You can even send and receive messages by means of computer "bulletin boards" or "E-mail," (the E stands for "electronic").

The telecommunications capability is probably not sufficient reason by itself to buy a computer unless, for example, you have a very large investment portfolio and want to be able to follow the market on your own computer. However, if you already have a PC or are going to buy one, it's surely worth looking into the telecommunications possibilities that now exist.

✓TIP   The single most important factor in deciding which personal computer to buy is determining what the primary use of the computer will be.

Here's how

BUYING A PC

First and foremost, it's terribly important to be realistic and honest with yourself. It might sound like a terrific idea to use the computer with a database management program so that you can organize all your files and retrieve information quickly and easily. But will you do it? If you're not a really organized person, it's unlikely that you would actually spend the 20 to 100 hours it would probably take to create a computer filing system.

The most popular PC application is word processing. So let's suppose you decide you need a computer for this purpose. Your first step should be to *explore word processing packages, not computers.* Start by reading the reviews of the different packages that regularly appear in computer magazines. Go back to older issues. Go to a computer store and ask for a demonstration of particular packages that you think are of interest. Talk to the salespeople in computer stores; they're often very knowledgeable. Ask friends for their recommendations. *Think long-term.* Features that enable you to create professional quality documents may turn out to be extremely desirable.

Once you've decided which word processing package is best for your real needs, and for those you think you may really have in the future, it's time to find out which computer will run that package. Each word processing program has its own memory and microprocessor requirements.

You're almost ready, but not quite, to begin your search for the right computer. Doing word processing also requires buying a printer so that you can produce typed documents. It also requires a computer display (monitor or screen) that gives you sharp, clear, easy-to-read letters.

Letter quality printers will give you type with the clean, sharp look of a good typewriter. *Dot matrix* printers, which are faster and generally less expensive than letter quality printers, produce letters that are made up of very closely spaced dots. The print looks good, but you can tell it was printed by a computer. At the other extreme, higher priced laser printers are fast, quiet, and produce print-quality documents that can offer a wide variety of printer fonts and can even include illustrations.

Small, portable computers come with the display built in, while for most desktop models you must purchase a separate monitor. Color monitors are nice to look at, are essential for graphics or games, and save eye strain. Different manufacturers offer different monitors. Only you and your eyes know what's best for you.

Now that you know the requirements of your word processing program, monitor, and printer, you can begin looking for computers that meet those minimum requirements. Other factors like versatility, price, portability, and manufacturer's reputation are also important considerations. On most, but not all, computers you can easily add more memory and special circuit boards that will increase your computer's versatility. Will you be moving the computer from room to room so that other family members can use it? Some computers are designed to be portable; others are not. Many computer manufacturers have gone out of business, leaving the owners of their equipment with no place to go in case problems develop. Buy a reputable brand of computer from a reputable dealer.

Prices of computers and printers vary widely but have decreased dramatically in recent years. It's best to "try before you buy." Try out several PC's that meet your requirements. See if you like the keyboard and the look of the equipment. All other things being equal, buy what you like best.

We used the example of buying a PC that was to be used mainly for word processing. Similar considerations are involved when purchasing a computer for other primary applications.

Don't let cost be the only determining factor. If you decide to buy a computer and you choose thoughtfully, you'll find it a worthwhile investment. Buy one that is capable of doing a little more than you now think you might realistically need so as to allow yourself some room in which to grow. Your increased confidence and expertise will surely warrant it.

INDEX